INTERTWINED

INTERTWINED

From Insects to Icebergs

Michael Gross

JOHNS HOPKINS UNIVERSITY PRESS | Baltimore

Johns Hopkins University Press
2715 North Charles Street
Baltimore, Maryland 21218
www.press.jhu.edu

Library of Congress Cataloging-in-Publication Data is available.
A catalog record for this book is available from the British Library.

ISBN 978-1-4214-4997-5 (hardcover)
ISBN 978-1-4214-4998-2 (ebook)

The author has adapted some material in this book from his previously published articles in *Current Biology,* including the following volumes and issues: vol. 24 (no. 24); vol. 25 (nos. 6, 12, 14, 16, 20, 23); vol. 26 (nos. 1, 5, 7, 10, 12, 14, 18, 19, 22); vol. 27 (nos. 2, 3, 5, 8, 13, 14, 16, 19, 22); vol. 28 (nos. 1, 2, 6, 11, 16, 22); vol. 29 (nos. 1, 6, 7, 12, 13, 21, 22, 24); vol. 30 (nos. 1, 5, 6, 11, 12, 14, 16, 17, 23); vol. 31 (nos. 2, 10, 11, 14).

Special discounts are available for bulk purchases of this book. For more information, please contact Special Sales at specialsales@jh.edu.

CONTENTS

INTERTWINED

Everything Is Connected

"Only connect!" has been the motto of my online presence since 1996—a quote from the novel *Howards End* by E. M. Forster, first published in 1910. Its protagonist, Helen Schlegel, continues her imagined speech thus: "Only connect the prose and the passion, and both will be exalted, and human love will be seen at its height. Live in fragments no longer." In those words, I recognized my budding attempt to bridge the worlds of science and literature as a science writer and my pleasure in discovering connections between scientific disciplines, and even between science and culture, that other scientists and writers may not have noticed.

I have found and made many connections over the years and have broadened my interests from the physical sciences into ecology and environmental issues. Increasingly, I became aware that ecology is all about how, in the living world, everything is connected to everything else. And this web of connections operates on a vast range of different scales, from molecular interactions (and thus to my background in biochemistry and the nanoworld) to global cycles of important chemical elements like nitrogen and carbon.

In between they include not only the predator-prey relations of the food web but also multiple ways in which species shape their environment and create opportunities for others. As ecology is a relatively young concept (the term was coined by Ernst Haeckel in 1866), many of these crucial connections are only beginning to be explored by science, although human activities have already started to destroy them. For instance, whales, sea birds, migrating fish (like salmon), and bears are all part of a global pump that transports nutrients uphill, against the flow dictated by gravity and the hydrological cycle. By the time scientists discovered this connection, its capacity to cycle nutrients was already reduced to a fraction of its former size.

Using numerous examples based on recent research and conservation issues around a broad range of ecosystems, this book shows us the connectedness of life on Earth and the many ways in which human activities have unwittingly disturbed these connections. We started doing so when we began to hunt with long-distance weapons, from spears to guns, catapulting ourselves to the top of the food web. We have carried on killing the wrong kinds of animals ever since, as we shall see in chapter 5.

Even by accidental interventions, such as carrying rodents on our ships, or by well-meaning ones, such as introducing raccoons to Germany in a misguided bid to enrich the fauna, we have messed up the biosphere considerably. Industrial-scale damage is done in the name of efficiency. Growing a single crop on vast surfaces is an efficient way to produce food, but it is the opposite of a functional ecosystem and is susceptible to large-scale pest invasions. By making agriculture more efficient, we have disrupted the checks and balances that are inherent in ecological networks, and we have tried to fix that by adding chemical pesticides and fertilizers to the mix. Only now, faced with the challenges of climate change, do we see a growing awareness of the benefits of producing crops in more diverse settings.

All of this isn't to say that change is bad. The complex web of life on our planet evolves, so it changes all the time, but typically on a much slower scale than the changes that we are causing. Incidentally, evolution presents us with a second set of connections linking all living things—they are all part of the big family tree of life, going back to a shared ancestor that first used RNA as genetic material (DNA and proteins are later inventions). Like ecology, evolution is a child of the nineteenth century, and understanding it is still a work in progress. In the last two decades, the opportunity arose to apply whole-genome analysis on a massive scale, to the extent that there is now a project underway to sequence the genomes of all eukaryotic species known to science. This will again redefine our understanding of how all living things are connected to each other through their family tree.

Evolution, for all its merits, is a very slow actor, and it can't adjust to all the changes that we are inflicting on our world—even though it does produce some adaptations, as we shall see in the context of urban ecology and evolution. Even climate change, which we perceive as an incredibly slow and gradual transition, is fast on a geological timescale and too fast for many ecosystems to adapt to by evolutionary processes.

Much of the damage we are causing can be attributed to simple recklessness, like overexploiting limited natural resources. However, another big part of it, I believe, is due to a lack of knowledge about all the invisible connections that tie life on Earth together. Consumers may, in one way or another, help to finance deforestation in the Amazon without realizing that they are also breaking the water pump that produces the rain essential to beef production in the Argentinian Pampas, which causes environmental problems of its own.

The concept of ecosystem services—the benefits provided to humans by healthy environments—has recently been explored as a way of making such crucial connections more visible. A key problem with our exploitation of natural resources is that the value they are adding to our economy isn't showing up in the balances. It is taken for granted while the natural systems work. We often realize too late—only once we have destroyed them and the services have disappeared.

Thus I believe it is extremely important to better understand, and spread the understanding of, all the connections that keep the wheels of the biosphere turning and thereby feed not only wildlife but also eight billion humans. We may be on track for a sixth mass extinction as well as a collapse of our civilization because of the damage we have caused already, but a better understanding could help us to limit the damage and learn to adapt to the dramatically different world we are creating in this new era, the Anthropocene.

In this spirit, I have vastly improved my own understanding of these connections while writing about ecology over the last 20 years or so, mostly for the front pages of the academic journal *Current Biology*. I'm presenting some of the lessons I learned recently in this book, necessarily in a linear order that doesn't do justice to the multidimensional web of connections I'm talking about. So feel free to pick and mix or jump around following the cross-references; it's all good, because, as I keep saying, everything is connected to everything else. Through sharing what I have learned, I hope to further spread understanding and make a small contribution to the effort to save the ecological connections that keep us alive.

CHAPTER 1

Plants and Their Little Helpers

Life is very nearly everywhere on the planet, from the deep sea to the high mountains, from the tropical rainforests of the Amazon to the rocks and subglacial lakes of Antarctica. For most of its four-billion-year history, however, Earth hosted life only in the oceans. The conquest of the continental surfaces was made possible by plants and the way they use photosynthesis.

Plants ultimately feed us and most life on Earth by turning inorganic carbon and solar energy into what we call food. Even the oxygen we use to "burn" carbs comes from plants. And we return the favor when we exhale carbon dioxide, which the plants can use again. These interlocking cycles of plant and animal metabolisms are the fundamental motor of multicellular life.

In keeping up their side of this vital exchange, plants don't act alone. The importance of microbes and fungi, not only in current biomass production by plants but also in the evolutionary processes that enabled plants to make the continents habitable, has long been underestimated.

More and more it looks like plants are crucially dependent on their microbiota—as are humans. Like us, carrying bacteria in our digestive systems, plants are holobionts—living systems based on the co-operation of multiple species. This insight has important implications, ranging from our understanding of deep evolutionary history to tomorrow's food security.

A billion-year success story

On Antarctica's bare rocks, lichens have been identified as far inland as latitude 86° south. These survive freezing, darkness, and drought, as long as some mild summer days let the temperature rise above zero and provide some melt water or vapor. The secret recipe for surviving under some of the most extreme conditions found on our planet lies in the symbiotic relationship between a fungus (mycobiont) and a photosynthetic organism (photobiont), which may be an alga or a cyanobacterium.

More than 400 species have been detected on the continent, with community composition depending on the prevailing weather as well as the presence or absence of seabirds. In the absence of guano, the cyanobacteria's ability to fixate nitrogen enables lichens to live entirely from the air, using the rocky substrate only for support. Elsewhere, lichens are a key indicator species for forest ecosystems, but they are also present in towns and cities, where they may grow on stone walls and roof tiles and often provide a charming patina to old stone buildings.

Given lichens' ability to appear among the first colonizers of such inhospitable habitats, it has been suggested that they may have been pioneers of the move from the oceans onto dry land. They could have played an important role in weathering minerals and thus preparing the soil for vascular plants to grow on. Recent molecular phylogenetic analyses of the fungi involved and their photosynthetic partners suggest, however, that the evolutionary history of lichens is shorter than anticipated but also more complex.

Both fungi and photosynthetic microbes have been around for more than a billion years. The ancestors of today's cyanobacteria began two-step photosynthesis around two billion years ago, setting off the biological production of oxygen. Early eukaryotes engulfed those photosynthetic bacteria and evolved to become algae and plants. The first fungi may have split from the lineage that eventually led to animals at least one billion years ago. But when did fungi team up with either cyanobacteria or algae (as they do today with both) to form a symbiotic relationship?

Dating the origins of lichens is important for our general understanding of the evolution of life on Earth, as lichens could conceivably have prepared the ground for vascular plants to expand on the continents around 440 million years ago. With their roots, vascular plants could substantially modify the land surface and make it habitable for other species. The key question, however, is, Did lichens make the land habitable for plants? Scientists since the times of Darwin have thought, based on the pioneering role that lichen take today in inhospitable environments like Antarctica, that they may have been the first complex life on land.

However, the fossil evidence to back up such speculation hasn't been found yet. Identifying early lichens from the fossil record is extremely difficult, and uncontroversial examples of early lichens are rare. A suggested "lichen-like" assemblage of filaments with algae that was described in 2005 and dated to 600 million years ago is not generally accepted as evidence of lichen. Subsequent molecular studies have begun to hint at younger ages of individual groups of lichen fungi. However, interpreting this information is difficult, because several distantly related groups of fungi form lichens, raising the questions, How many times did lichenization evolve, and how old were these transitions?

In a bid to clarify when lichens originated and vegetation first arrived on land, Matthew Nelsen at the Field Museum in Chicago and colleagues calculated age estimates for phylogenetic trees, including most groups of present-day lichen-forming fungi and algae.[1] Because of the lack of conclusive fossil evidence, the researchers "used an alternate approach in which [they] placed molecular phylogenies of fungi and algae in a temporal framework by inferring the evolutionary relationships of extant taxa and linking them with fossil data." Nelsen and his colleagues "then inferred where and when in the fungal phylogeny lichenization evolved by using a process known as ancestral state reconstruction. The ages of the nodes reconstructed as lichenized could then be placed in the context of land plant and terrestrial ecosystem evolution."[2]

Fungal and algal family trees provided independent data sets, which the researchers could then compare to see if they made sense. They "found that the ages of many algal lineages broadly coincided with those of the fungal clades associated with them."

The researchers' modeling found, with high confidence, that before 440 million years ago no ancestors of today's fungi arose with the ability to form lichen. Before that time no lichens were growing on Earth's bare land surfaces, which were covered by nothing more complex than bacterial mats and mosses, that could have shaped the environment and made it easier for subsequent plant invasions.

Similarly, the authors argue that the results obtained for algae and cyanobacteria are best compatible with an origin of lichens that is less than 440 million years ago—even though cyanobacteria as such are as old as oxygen-producing photosynthesis.

Overall, the most likely date for the origins of lichens would be around 250 million years ago—so one could speculate that they might have started spreading after the biggest mass extinction of all, which marked the end of the Permian. Other new arrivals spreading across the landmass of Pangaea in the new Triassic period (ca. 252–201 million years ago) included the first dinosaurs as well as the first mammals. However, the authors caution that there is considerable uncertainty in timing and exact locations within the fungal phylogeny.

In a more detailed analysis of the evolution of the largest group of lichen-forming fungi, the class Lecanoromycetes, Nelsen and colleagues found that the relationship between fungal and algal species has been more dynamic than anticipated, with lineages reverting from symbiont status to a solitary life.[3]

Although losses and switches of symbiont status had been known before, the researchers were surprised to find that this happened much deeper in the evolutionary past of the Lecanoromycetes. Modern-day non-lichen descendants of early lichen formers have largely adopted lifestyles as plant parasites and decomposers, but it was unclear which of these roles their ancestors adopted when they stopped forming lichens. According to Nelsen, "Some of these descendants have continued to persist in a non-lichenized state, while others picked up another algal partner, in some cases at roughly the same time as other groups of non-lichenized fungi were evolving lichen associations with that same group of algae."[4]

There were times and places in Earth's evolutionary history when the formation of a new symbiotic partnership as a lichen was favored by environmental conditions. Wet tropical forests dominated by flowering plants, which became widespread in the Cretaceous (145–66 million years ago), were one such environment favorable to algae and the fungi that might end up hooking up with them. Multiple lineages took up or regained the lichen habit during that time.

Comparing the diversification rate of the Lecanoromycetes to another group of fungi, the Agaricomycetes, which are mushroom-forming fungi, the authors found very different patterns. The Agaricomycetes expanded rapidly in the Jurassic (201–145 million years ago), while the Lecanoromycetes diversified more slowly and steadily.

As Jen-Pan Huang at the Field Museum and colleagues, including Nelsen, reported in a separate paper, the bulk of the diversity of present-day macrolichens—those with conspicuous plantlike structures that may look like leaves and branches—only diversified after the last mass extinction, 66 million years ago.[5] Thus, the current diversity and spread, with lichens dominating 7% of the land surface, is a relatively modern phenomenon in the history of life on Earth.

Even though the theory that lichens helped life to conquer the continents failed to stand up under scrutiny, recent research suggests that, more than 700 million years ago, a different co-operation also involving fungi may have played that role.

Susana Magallón from the National University of Mexico at Mexico City, with colleagues from the United States, Norway, and Germany, investigated the evolutionary history of the interactions of plants and fungi over the last one billion years.[6] Along with the symbiotic relationships observed in lichens, other ecological interactions to be considered include parasitism and saprotrophy, which is when fungi obtain their nutrients from the decomposition of plant material.

Magallón and her colleagues systematically compiled and analyzed the phylogenies of the kingdoms of plants and fungi separately, and then sought to establish how interactions between them may have influenced their evolution, as reflected in diversification rates, for instance. This investigation led the researchers back to the time, one billion years ago, at the beginning of the Neoproterozoic era, when life only existed in the oceans and multicellular organisms had not yet evolved.

According to these investigations, green algae were the first of the relevant groups to colonize land. Within the same era, some

720 million years ago, the ancestors of fungi lost their flagella, presumably because they no longer swam in water.

The Neoproterozoic ended with the famous explosion of multicellular biodiversity that we see in the Ediacaran (635–541 million years ago). In the Paleozoic era, which began 541 million years ago with the Cambrian, the shared ancestors of all today's land plants came ashore. Magallón and colleagues conclude from their evidence that this move was greatly helped by interactions with fungi. The important question is the timing: Which phylogeny diversified earlier, that of fungi or that of seed plants?

The authors found that the phylogenies of both seed plants and the fungi now associated with their roots converged to their respective common ancestors at a point around 720 million years ago, suggesting that both conquered dry land together, much like lichens can today colonize extremely inhospitable environments.

Specifically, the fungal division of Glomeromycotina (formerly known as Glomeromycota) is today associated with plant roots and rootlike structures forming arbuscular mycorrhiza, except for one species, *Geosiphon pyriformis*, which forms a lichen with a cyanobacterium. The earliest fossil evidence of Glomeromycotina dates to around 400 million years ago, in the Devonian, but the new analyses push the origins of this group back in time to a window between 715 and 606 million years ago, much closer to the beginnings of land plants.

Another clade of fungi, the Mucoromycotina, which has been seen associated with early seed plants, may have also played a role in enabling plants to expand on dry land. The authors also point out that fungi coming ashore may have interacted with other organisms, including algae, amoebae, and bacteria, before the cooperation with seed plants started to take off and changed the face of the planet. It is worth remembering that much of the land surface that we walk on and tend to regard as geology is made of the mortal remains of living organisms, which shaped the surface of the Earth beyond recognition, especially since the rise of terrestrial plants.

Another significant milestone in the shared history of plants and fungi occurred when lignophytes, early ancestors of the lineages that lead to today's flowering plants (angiosperms) and gymnosperms (conifers and others), developed the ability to grow wood fibers. This innovation in the Silurian, some 420 million years ago, required a new type of cell division and led to very resilient, lignin-rich plant materials. The tall structures and deep roots that plants can grow thanks to the development of the material we now call wood enabled them to grow away from swamps. They could spread across continents and create their own environments, the first forests. However, the decomposition of wood fibers challenges many organisms and even today's technology, for instance in the production of biofuels from plant waste.

This innovation might have become a dead end if lignin had remained indigestible for all living organisms. Fortunately, fungi evolved the ability to degrade lignocellulose and thereby enabled the recycling of nutrients from dead trees to allow the growth of new ones. This new ecological interaction helped both sides. It produced thriving forests that covered much of the land surface, and it led to a boom in fungal diversification, making the class of the Agaricomycetes the largest one now in existence.

Another important interaction between early plants and fungi happened invisibly below ground. Fungi started colonizing the roots of plants and engaging in their nutrient provision. These ectomycorrhizal fungi, traced back to origins just after the evolution of wood and deep roots, are now an important factor in the growth of many plant species, including cultivated crops. Like the fungal partners in lichens, these underground symbionts enable the plant to thrive in more challenging environmental conditions.

The ecological interactions that have enabled plants and fungi to make the continents habitable for themselves as well as for animals continue to play important roles in the present. They are indispensable for the ecological balance of the biosphere and for the food security of its human inhabitants.

Life on plants

Among the many different interactions between plants and fungi, the arbuscular mycorrhizae are the most widespread today. They are called arbuscular because they form finely branched hyphae within the cells of the plant root (in contrast to the ectomycorrhizae, which associate externally). More than 70% of current plant species, including many crop plants, rely on these fungal helpers for their nutrition.

Optimizing the work of root-associated fungi such as the arbuscular mycorrhizae can help to minimize the need for fertilizers in agriculture. However, their presence underground and functional co-operation with the plant roots is difficult for researchers to establish without harming the plant.

For the purpose of breeding and growing crop plants, it would be useful to know exactly what happens underground—for instance, whether a new sapling has successfully established a symbiosis with arbuscular mycorrhizae. This is why Ian Baldwin's group at the Max Planck Institute for Chemical Ecology in Jena, Germany, investigated the model plant *Nicotiana attenuata* (coyote tobacco, a wild tobacco species) to look for biomarkers that might report the status of the root symbionts without the need to damage the plant.[7]

Specifically, the researchers homed in on a group of metabolites known as blumenols, which are C-13 compounds arising from the decomposition of carotenoids and are known to be produced in the roots of the plant. Glycosides of the blumenol-C class of compounds, in particular, had been found in roots when these hosted arbuscular mycorrhizae.

Baldwin's group, together with several other teams, systematically looked for these compounds in *Nicotiana* plants with normal ability to associate with the fungi, as well as in mutants that lack an essential genetic trait and are therefore unable to host the fungi. Using advanced methods of liquid chromatography and mass spectrometry, the researchers could identify five different compounds deriving from blumenol-C that are present in roots only when they host arbuscular mycorrhizae.

Extending the search to stems and leaves, Baldwin and colleagues found that two of these compounds are also detectable above ground. While their concentration is much lower above ground than in the roots, the compounds are still readily detectable with state-of-the-art mass spectrometry. Further investigations and control experiments confirmed that these markers are suitable for a specific and quantitative assessment of symbiotic activity in the roots. Tests with other plants not closely related to the *Nicotiana* genus also showed that these markers appear to be widespread among plants, so the method is likely to be widely applicable.

This discovery creates opportunities to use high-throughput experiments and field studies to examine the symbiotic relationship without harming the plant. Plant breeders could use it to develop new variants that can use their symbionts more efficiently and therefore require less phosphate fertilizer or thrive better under environmental stress.

The interactions that plants get involved in below ground, in the rhizosphere, are typically invisible to us and often taken for granted. If the same kinds of plants have been growing on the same soil for many years, one assumes that the soil microbiota is healthy—as long as the plants are healthy. Still, it is worth knowing what a healthy microbiome is, to safeguard plant health in agriculture and to facilitate restoration of natural plant communities.

Roberta Fulthorpe from the University of Toronto Scarborough, Canada, and colleagues studied the microbiota from coffee plants (*Coffea arabica*) sampled in several locations across Central America with a wide range of environmental conditions to establish a core microbiome for the coffee plant.[8] Although the composition of the rhizosphere communities varied between sites, the researchers were able to identify 26 bacterial and 31 fungal species that were consistently present across all conditions and thus met the criteria to be included in the core microbiome. In general, they found the composition of the bacterial community more consistent than that of the fungal one. Some of the species are already known to be beneficial, but more research is needed to establish what their role is in maintaining the health of the plants.

The existence of a core microbiome suggests that these populations are to an extent controlled by selective factors specific to the plant and that this microbiome could be optimized further.

With the idea of improvements for a globally important crop in mind, Brooke Bissinger's group at AgBiome in North Carolina studied the microbiomes of sweet potato (*Ipomoea batatas*) plants at two separate farm sites in North Carolina.[9] They, too, found differences between the sites but were able to identify a core microbiome specific to healthy plants of this species. The work ultimately aims to make sweet potatoes grown in the developing world more resilient to insect threats.

What can go wrong if the microbiome stops supporting the plant is illustrated by the *Verticillium* wilt disease that is threatening olive plantations across the Mediterranean region and also affects other crop plants. It is caused by a fungus called *Verticillium dahliae*. If it is present in the soil, the fungus infects root systems and blocks the vascular systems until the plant dies. The overall effect produced by the disease above ground looks like the result of a severe drought.

Carlos Garay from the Institute for Integrative Systems Biology in Valencia, Spain, and colleagues analyzed the mixture of messenger RNAs (called the metatranscriptome) found in samples from an earlier study that were obtained from leaves and roots of olive trees affected by the disease.[10] They found that the attack of the *Verticillium* fungus on the root system is supported by a complex network of other species, including bacteria, other fungi, and amoebae. In the later stages of the disease, even the plant's symbiont helpers turn against it and accelerate its demise.

For the parts of the plant that grow above ground (the phyllosphere), researchers are still figuring out who is a friend and who is a foe of the plant. Bacteria found on leaves, for instance, may just be present by coincidence, they may be harmful, or they may help the plant to defend itself against the harmful bacteria.

The advantage of the phyllosphere is that its microbial population can be more readily controlled and manipulated than the rhizosphere. Britt Koskella's team at the University of California, Berkeley, studies the leaf microbiome of tomato plants. In a report published in 2018, for instance, Maureen Berg and Koskella investigated whether a leaf wash containing microbes from healthy field-grown tomato plants could protect the lab-grown plants from the effects of a subsequent experimental infection with the bacterium *Pseudomonas syringae* pathovar tomato (Pst).[11]

The results confirmed that the healthy leaf microbiome does have a protective effect, but some of the details were surprising. The researchers observed that a higher dose of the bacterial treatment (leaves that received more microbes) offered less protection than a lower dose. While the reasons are still unclear, this complexity suggests that the application of probiotics for plants in agriculture needs to be tested carefully.

Moreover, fertilizing the plants before the experiment abolished the protective effect, which, according to Koskella, "raises many important questions about common agricultural practices that may be substantially altering the relationship between plants and their microbiomes."[12] This effect, too, needs further investigation and optimization of protocols for application.

In a follow-up work, Koskella's group could achieve the same protection against Pst by applying a microbial inoculum to tomato seeds before planting them.[13] They identified two bacterial species, *Pantoea agglomerans* and *Pantoea dispersa*, which were sufficient to achieve the protective effect, but again they found evidence that more microbes are not necessarily better.

Koskella and her colleagues then went on to investigate how the leaf microbiome on tomato plants can, over several rounds of passaging from one group of plants to the next, adapt to the host and its growth conditions.[14] They started by inoculating their lab-grown tomato plants with leaf microbiota gathered from field-grown plants. After cultivating for eight weeks, they collected the microbiome from the leaves, analyzed its composition, and used it to inoculate a fresh group of tomato plants. They repeated this four times, so they could study five generations of tomato-leaf microbiome in the laboratory.

Their analyses showed that over the generations, the diversity in the microbiota reduced, with only a quarter of the taxa from the original field sample still detectable in the last round. This suggests that non-adapted microbes, which may have been present in the field sample by accident, were displaced by a well-adapted core microbiome that became more and more stable. The advantage to the plant is that such a well-adapted and stable microbiome can fend off other kinds of environmental microbes that could be harmful.

To test the microbiome's protective effect, the researchers treated a fresh group of plants with a mixture of the microbiota sampled from the last round of the experiment and those from the first round. They found that the adapted bacteria again pushed the others out, suggesting that this optimized leaf microbiome had become resistant to microbial invaders from the environment.

But how does a plant cultivate its protective microbiome? The groups of Sheng Yang He at Michigan State University and Xiu-Fang Xin at the Chinese Academy of Sciences in Shanghai discovered a gene network in *Arabidopsis* that is necessary for the maintenance of a healthy population of microbes within the hollow spaces of leaves.[15] In previous research studying the genetic foundations of *Arabidopsis*'s susceptibility to pathogenic bacteria, the researchers noticed that one quadruple mutant of *Arabidopsis* had an abnormal population of endophytes and showed signs of disease while the epiphytes, the microbes living on the leaves, were normal. The mutations in this strain disrupt the plant's immune system as well as its control of hydration levels within the leaves.

To strictly prove the causal connection between the disrupted gene network, the composition of the microbiome, and plant health, He's group developed a germ-free growth microchamber for their mutant and wild-type plants. Starting from sterile leaves, they could add tightly controlled populations of microbes and observe their effects.

In their comprehensive study, He, Xin, and colleagues established that the plant's health depends on both the genetic setup and the healthy set of endophytic microbes. In the mutant, even an inoculation with the good germs doesn't restore health. Conversely, transplanting a sick microbiome to a plant with the intact genes can make it sick. The authors compare the resulting plant disease to irritable bowel syndrome in humans and describe it as a dysbiosis of the phyllosphere. The researchers expressed their hope that this work will ultimately help to feed the world.

Out in the fields, plants not only have a wider range of microbes to contend with but also the all-important insects, which add to the complexity of the ecological network. Parris Humphrey at Harvard University and Noah Whiteman at the University of California, Berkeley, characterized the three-way interaction between a plant (*Cardamine cordifolia*, or bittercress; family Brassicaceae), its leaf microbiome, and an insect herbivore (*Scaptomyza nigrita*, or the common leaf-mining fly) and discovered both an increase and a composition shift in the leaf microbiome in plants affected by the herbivore.[16] The shift favored *Pseudomonas syringae* strains that may or may not be pathogenic to the plant. Wenke Smets and Britt Koskella discussed this work in terms of herbivory-inducing dysbiosis in the plant leaves.[17]

In another example of ternary ecological interactions happening on plant leaves, Sybille Unsicker's group at the Max Planck Institute for Chemical Ecology in Jena, Germany, studied the development of gypsy moth caterpillars (*Lymantria dispar*) on poplar leaves (*Populus nigra*) infected by the rust fungus *Melampsora larici-populina* based on the observation that infected trees are more likely to be attacked by the insect larvae.[18] The researchers established that the caterpillars are attracted to infected leaves and prefer these over uninfected leaves because they detect the sugar alcohol mannitol emitted by the fungi. As the caterpillars develop more quickly on infected leaves, the authors conclude that at least the young larvae consume fungal spores in addition to leaf material. Nutritional benefits include greater levels of total nitrogen, essential amino acids, and B vitamins present in fungal tissues.

Life on plants, with its multiple interactions across kingdoms, is an important factor determining the life of the plants hosting it. Our management of plant life from agriculture—essential for our day-to-day survival—through to conservation—essential for the survival of a habitable planet in the long term—should be based on a better understanding of these complex networks, but we are only beginning to learn how these microbiotas originate, how they evolve over time and respond to changing conditions, and how plants support and manage them.

How plants grow their microbiome

Plants host significant microbiota in four different habitats, as microbes can be associated above or below ground and in each case outside or within the plant's tissues. In technical terms, the four options are the leaf endosphere, leaf surface, root endosphere, and rhizosphere. Which microbes are found in each is expected to depend on influences both from the environment (including other plants) and the plant itself, and it likely changes over time, as seasons change and the plant grows.

Melissa Cregger's group at the University of Tennessee, Knoxville, addressed these complexities with an experimental study of poplar seedlings. The researchers followed the initial assembly of the poplar microbiota in all four plant-associated habitats, using 10 genotypes from two different poplar species, *Populus deltoides* and *P. trichocarpa*.[19] They sampled and sequenced microbes and fungi at three points spread out through the first growing season.

From these analyses, the researchers found that across all four habitats, the microbiome composition changed dramatically within a few months. Differences observed across time on the same plant were generally more significant than those seen at the same time between plants of different genotypes or even species. For archaea and bacteria, the changes led to the selection of a smaller group of species, as the host plants established their connection with the desirable microbes.

Some specific findings of species trends differ from those observed previously with other kinds of plants. Thus, the study found Gammaproteobacteria gradually replacing Alphaproteobacteria in aboveground habitats, which differs from previous studies with grass species. Below ground, previous research with other plants, including *Arabidopsis* and *Citrus* species, had suggested that the increasing exudation of plant chemicals into the soil favors Proteobacteria and Bacteroidetes over Actinobacteria. The poplar study, by contrast, found Actinobacteria on the rise. Overall, the differences developing over time were stronger in the exosphere than in the endosphere—that is, on leaves and around roots compared to within the plant structures.

For fungi, however, the authors obtained very different results, implying that selection was weak and dispersal factors played a bigger role. Further complexities arose as attempts at tracking the sources yielded mixed results and core microbiomes identified for each habitat were consistent but didn't necessarily become more dominant over time.

The relationship between the poplar hosts and their microbial and fungal guests is a complex one from the start, with many competing influences to consider. Time, the environment, and stochastic influences appear to play a stronger role than the genotype of each plant. However, the authors speculate that as the trees grow and mature over the years, their genotype may get a stronger role in shaping the specific character of their microbiome, which is observed in mature trees.

Young plants may recruit their microbiome from the environment but also from other plants around them. These plants may be the same or different species, and they may be the same age or older ones with more established microbiomes.

Geneviève Lajoie and Steven Kembel, from the University of Quebec in Montreal, Canada, conducted a large observational study of community effects on the leaf microbiomes with 33 tree species along an ecological gradient from deciduous to boreal forest in eastern North America. They found that the species composition of the leaf microbiomes was influenced not only by the identity of the specific host but also by its neighbors. Specifically, their data suggested that the less abundant tree species are often swamped with the microbes dispersed by their majority neighbors, making it more difficult for them to maintain their separate, species-specific microbiome.[20]

Addressing such neighborhood effects with a controlled experimental setup, Kyle Meyer and colleagues, working with Britt Koskella, manipulated the neighborhood interactions between tomato (*Solanum lycopersicum*), pepper (*Capsicum annuum*), and bean (*Phaseolus vulgaris*) plants.[21] The researchers cultivated the plant of interest with neighbors of the same or different species or none at all. After 30 days, the focal plant was removed and replaced, while the surrounding plants remained in place and continued growing. This way, the parameter of age and biomass of the surrounding plants was introduced as an additional variable.

The analyses of bacterial communities in the leaf microbiomes of the focal hosts showed that their initial success in selecting their specific microbiome decreased as the biomass of surrounding plants increased. As larger plants offer larger surfaces for microbes to settle on, there are larger source populations for dispersal that can then invade the younger plant studied and overwhelm their filtering efforts. Among the species investigated, pepper plants were vulnerable to invasion from older neighbors at an earlier stage in neighborhood development than tomato and bean plants, possibly because pepper plants naturally have a less abundant leaf microbiome.

To directly study the host's filtering ability under controlled dispersal conditions, Meyer and colleagues conducted cross-inoculation experiments. In these controlled conditions, even a second-hand neighbor effect could be seen, in that the result observed on the recipient plant reflected the neighbors of the donor plant, which the authors interpret as evidence of previous transmission that still lingers in the present microbiome.

Both studies show that while each plant is programmed to cultivate its own microbiome matching its species and growth conditions, the neighborhood of other plants, especially those of different species and of more advanced maturity, can lead the microbiome assembly in different directions. These influences can become important in real-world ecological challenges, such as reforestation or dealing with invasive species.

Host filtering and selective support of microbiota is especially important where some of the species involved have the potential to become pathogenic. The model plant *Arabidopsis thaliana* does not have the setup to form symbiotic structures but still gets settled by a variety of fungal endophytes. Among the fungi associated with healthy plant roots in the wild, researchers at the Max Planck Institute for Plant Breeding Research in Cologne, Germany, found that many were closely related to pathogenic species and have retained pathogenic traits.

Fantin Mesny and colleagues sequenced 41 of such fungal isolates from the wild and found pathogenic traits, including numerous cell-wall digesting enzymes.[22] When they tested these fungi in recolonization experiments with *A. thaliana* plants, the researchers found a wide range of effects, from beneficial to detrimental. The highest root-colonizing potential often coincided with the highest risk of damaging the host plant, while those species that were proven beneficial in one-to-one association experiments were less successful colonizers and therefore also less abundant in natural populations. Previous work from the institute suggests that, in a more natural context, the plant's immune system and bacterial microbiome co-operate to keep the negative effects of aggressively colonizing fungi under control and to enable the plant to reap the beneficial effects.

Protists as predators of plant-associated bacteria and fungi are another factor shaping the fate of the whole system, the holobiont.

Their role has been neglected so far, partly because they are difficult to grow in culture. To amend this gap in research, Kenneth Dumack at the University of Cologne, Germany, and colleagues have established a culture collection of protists associated with *Arabidopsis thaliana*, which are freely available to researchers around the world.[23]

There is no shortage of examples of resident microbes protecting their plant hosts from disease, but it remains difficult to make generalizations or to assess how widespread the protective effect is and to predict how it could be applied.

In an effort to systematically assess the protective potential of bacteria in the leaf microbiome, Julia Vorholt's group at ETH Zurich, Switzerland, tested all 224 *A. thaliana* leaf isolates of the collection At-LSPHERE for activity against the pathogen *Pseudomonas syringae* pv. tomato DC3000.[24] Although previous in vitro experiments had only detected activity against this pathogen in two of the strains, the screening on plants revealed more than 10% of strains offering full protection against disease, another 10% showing partial protection, and the remaining 80% causing no marked reduction in disease symptoms.

Many of the protective strains were part of the core group of taxa often observed in the phyllosphere. Thus the authors expect that strains' protective effect would normally be present in a typical natural environment.

The mechanisms of the protection can be different, ranging from direct attack on the competing bacteria to indirect effects, such as competition for nutrients or stimulating the plant's defenses. To study intrinsic mechanisms, the researchers compared the genomes of a range of strains from the same family offering different levels of protection. They found an association between certain gene clusters and protective effects. Most of these clusters were predicted to be associated with type VI secretion system

(T6SS) components. However, a mutant with inactivated T6SS still showed a low level of protection, suggesting that several mechanisms are working in combination.

As we are learning about the importance of the microbiota for plant health and pondering ways of using this knowledge, it emerges that some pathogens are already ahead of us. We have noted above that the pathogenic fungus *Verticillium dahliae*, which causes wilt disease in crop plants including olive, tomato, and lettuce, can turn beneficial microbiota against host plants. In line with that observation, Nick Snelders from the University of Cologne, Germany, and colleagues reported that this fungus secretes an antimicrobial protein as an effector specifically targeting the microbiome of host plants to promote infection.[25]

Beyond providing protection from disease, resident microbiota may also help plants to cope with environmental stress. This aspect is becoming more important as climate change threatens to affect the environmental stability of crops and wild plants alike. An important question is whether and how, as plants are forced to adapt to different environments, their microbial guests will evolve as required for the new conditions. Related to this is the question whether any intervention in the microbiome can make plants more resilient to climate change and environmental stress.

One group of species that has experienced both environmental change and human interventions is the domesticated apple tree (*Malus domestica*), together with its wild ancestors and relatives. Ahmed Abdelfattah from Graz University of Technology, Austria, and colleagues chose to study the microbiomes of several *Malus* species to assess how the microbiota changed through domestication and translocation, and whether they co-evolved with the host plant.[26] Apple trees were first domesticated from *Malus sieversii* in central Asia; they expanded westward along the Silk Road and hybridized with other *Malus* species on the way.

The researchers found that, unlike other crop species, the domesticated apple tree has a more diverse and abundant microbiome than its relatives in the wild. Phylogenetic relations between the microbiota broadly reflect those between the host plants, suggesting that, as apple trees were bred and transported to new environments, the resident microbiome co-evolved with the host.

This finding appears to suggest that, as climate change shifts vegetation zones and plants will have to cope with new environments, a concerted adaptation of the holobiont is possible, at least on the timescale of the domestication of crops, and without conscious human manipulation. Whether it can be achieved on the more rapid timescale of the climate catastrophe, and with human interference, is a different question. All the research into plants and their microbiomes has shown that these are extremely complex networks of interactions. To be able to guide them in a desirable direction, we would have to understand them much better.

Could plants have cognitive abilities?

When plants exchange chemicals with their microbiota, with herbivores, or with each other, the processes are typically described in terms of feeding on nutrients or using toxins as chemical weapons. On a slightly more sophisticated level, the chemicals could just carry information that has no nutritional or toxic element, like insects do, for instance, with their pheromones (external hormones). And if plants transmit information, can they store information too? Does the whole add up to a cognitive skill set? This idea sounds slightly absurd, so we might as well consult Lewis Carroll's famous compendium of absurdity, *Alice's Adventures in Wonderland*.

During her adventures in Wonderland, Alice encounters a blue caterpillar smoking a hookah and responding to her questions with very unhelpful answers and counterquestions. With the caterpillar's conversation style, Lewis Carroll is said to have been mocking his colleagues at Oxford University, but the pipe smoking is more mysterious. Why would an insect smoke, and how would it even be able to survive this activity, given that plants produce nicotine precisely because it is a potent toxin for insects?

As it happens, Carroll's vague description of the blue caterpillar fits the tobacco hornworm, the larval stage of *Manduca sexta*, which feeds on tobacco plants and is remarkably resistant to nicotine. When kept in the laboratory and fed with wheat germ, the caterpillar turns blue due to the lack of plant pigments in this diet. This fact would have been known to naturalists among Carroll's contemporaries.

While Carroll describes the caterpillar as sitting on a mushroom cap and enjoying his tobacco via a hookah, it is more interesting to consider what happens when a caterpillar gets its tobacco fix directly from the plant. *Nicotiana* has the ability to respond to herbivore attacks by increasing nicotine production. If the attacker is *Manduca sexta*, however, the plant recognizes the chemical composition of its saliva and appears to "know" that pumping out more nicotine doesn't help its cause. Accordingly, it refrains from expending energy on producing the toxin. While the attack of a tobacco hornworm can devastate an entire plant, the ecological picture is more complicated, as the adult stage of this species is a pollinator of the tobacco plant. The individual plant, however, has another defense mechanism in reserve, as the chemicals released by damaged leaves may attract predatory mites that attack the caterpillar—a case of tritrophic interaction, where the prey attracts the enemy of its predator.

But does it make sense to use cognitive vocabulary in the context of plants? Do they know which insect is nibbling on their leaves, remember past threats, communicate threats to conspecifics, or call for the help of a third party? Until the early twenty-first century, such ideas were the domain of esoteric philosophies and considered to be close to pure fantasy, like Carroll's famous tale and its talking, smoking caterpillar. A deeper understanding of the chemical signaling of plants above and below ground is beginning to make the idea of plant cognition more respectable.

Plants cannot speak, and it is unlikely that they can listen when overenthusiastic gardeners speak to them, but since the 1980s, chemical communication channels have been discovered at an increasing rate, with a growing number of recipient species found to tune in. In the beginning, in 1983, David Rhoades from the University of Washington in Seattle reported that willow trees can gain resistance in the neighborhood of conspecifics damaged by herbivores and speculated that an airborne signal molecule acted as a warning—a hypothesis that was confirmed later that same year by workers studying poplar trees.[27]

When a plant sends out volatile molecules indicating that it is being attacked by herbivorous insects, this is a piece of public information that could be received by various interested parties, including not just other plants but also herbivores and indeed carnivores that may want to attack the feeding herbivores. Additional recipients of the signals were identified over the years, including predatory mites (in 1988), parasitoid wasps (1990), predatory bugs (1995), ladybirds and moths (2001), nematode worms (2005), parasitic plants and anatomically remote parts of the emitting plant (2006), and birds (2008).

Which of these species is the "intended" recipient of the signal remains to be established. While there might conceivably be an evolutionary benefit to be gained if closely related conspecifics are efficiently warned of danger, or if predators are attracted to the grazing herbivores, the relatively short reach of the chemical signal supports the hypothesis that the signaling evolved as a communication between leaves of the same plant. These may be nearby in space, and thus subject to a shared risk from herbivore attack, but still distant in the branching structure, and thus difficult or impossible to reach via the internal fluid channels of the plant.

Richard Karban from the University of California, Davis, and colleagues reported a detailed investigation of the specific volatile profiles of individual plants of sagebrush (*Artemisia tridentata*). These authors found that there are characteristic and heritable differences in the chemotype of the plants' warning signals. Warnings received from plants with the same or similar chemotype offer more efficient protection, which suggests that these chemotypes represent a kind of kin recognition in plants.[28]

Even though hundreds of herbivore-induced volatiles have been identified, precise information on their concentration in the air surrounding the emitting plant and their further fate in the environment remain incompletely understood. Most importantly, the receptor mechanism, the "nose" of the plant being warned of the danger, has remained elusive. There are indications that the accumulation of volatiles in the plant membranes may play a role and that epigenetic mechanisms may enable this information to be stored and passed on to the next generation, but just how a plant sniffs out danger remains to be discovered.

An improved understanding of these mechanisms might lead to new approaches in sustainable pest control—one of the reasons why this area is under intense investigation.

One highly specialized and relatively well-studied area of chemical communication by plants is the pollination strategy of sexual deception. Orchids on several continents have independently evolved the ability to mimic the pheromones of female insects to attract males. This deception is so successful that copulation attempts are frequent, and the insect's misdirected energy secures the critical step of pollination without yielding any reward for the pollinator.

Researchers are still uncertain as to how this mischievous streak evolved or why it has evolved repeatedly. A preferred hypothesis in many cases is that of pre-adaptation, whereby the chemicals involved were co-opted from other functions and re-assigned, with modifications, to the task of fooling the pollinators.

However, in their investigations into the semiochemicals used by Australian sexually deceptive orchids, Rod Peakall's group at the Australian National University in Canberra kept uncovering novel compounds. For example, in *Drakaea glyptodon*, alkylpyrazines and a novel hydroxymethylpyrazine are the signaling molecules that the orchid uses to trick its pollinator.[29] The flowers have also evolved a labellum (lip-shaped petal) that mimics the shape of the flightless female wasps. In closely related *Chiloglottis* orchids, specific blends of unique compounds called chiloglottones are used to attract the males of only one species of thynnine wasp per orchid. Many more cases of chemical deception are yet to be analyzed in detail.

So far, pyrazines in plants are extremely rare, while chiloglottones are not known elsewhere, and no other functions in plants have been established. Therefore, the authors hypothesized that this may be a case where a chemical deception evolved via an evolutionary novelty, rather than pre-adaptation. The authors acknowledge, however, that proving the alternative explanation requires a comprehensive knowledge of the molecular and genetic basis of the production of these compounds. Further, an in-depth study across a range of species is required to trace the molecular phylogeny to the point where the chemical trickery originated.

Birgit Oelschlägel from the Technical University of Dresden, in Germany, and others have uncovered the chemical deception that one flower from the genus *Aristolochia* uses to attract flies.[30] This is the genus that produces some of the largest known flowers and some that famously smell of rotting flesh—all in the cause of deceiving flies. In the case of the species studied, *Aristolochia rotunda*, the volatiles are barely perceptible to the human nose, but Oelschlägel and colleagues showed that the plant has a very specific smell to its pollinator species, the kleptoparasitic fly Chloropidae. These flies feed on secretions from insects being killed and eaten by spiders. *Aristolochia rotunda* mimics the odor of these secretions to attract the flies.

The system described is unusual in that the plant mimics the chemicals emitted by dead insects, sending out the false promise of a meal for the flies, rather than mimicking the chemicals of live conspecifics signaling sexual opportunity. In contrast with bees, the flies hope to find food for themselves, not their offspring.

As more cases of chemical communication between plants and pollinators emerge, it becomes clear that the mix of chemicals tends to be highly specific to the species pair. This suggests that the ability to emit these specific chemical signals may have played a role in the speciation of the plants involved.

Volatile signaling may be difficult to analyze, but another communication channel important for a plant's interaction with its environment is literally hidden from view—its root system. The importance of the underground space around the plant roots, populated by a wide range of species attracted by chemicals secreted from the roots, was first recognized by Lorenz Hiltner (1862–1923), who coined the term "rhizosphere" in 1904. More than a century later, the concept has emerged as one of the most significant areas of plant science.

Below ground, as above, plants interact with their own kind and with multiple other species in complex ways, as we have seen earlier in this chapter. It has been shown that plants spend significant parts of their metabolic energy on substances that serve to feed the rhizosphere microbiome, which in return provides valuable services that may improve the plant's growth, development, nutrition, or immunity. This exchange of goods is so rich and complex that economic theories have been applied to characterize it.

As the success of the interaction with the microbiome depends on the genotype of the plant, there is concern that breeding for agriculture may have reduced the suitability of crop plants for some of the underground interactions whose benefits may not be immediately obvious to breeders, such as immunity. A better understanding of the whole rhizosphere could thus help to restore the natural defenses of crop plants and enable a more sustainable type of agriculture.

Progress is being made in the analysis of the chemical nature of the communication in the rhizosphere. Even small plants like *Arabidopsis thaliana* exude more than 100 different compounds into the soil. Sampling has to be well-considered to ensure that analyses are representative of field conditions, and the most sophisticated methods of metabolomic analytics are required to gain quantitative insights into the complex and dynamic situation around the roots of plants.

Direct investigation of signaling in the rhizosphere is currently focused on communication between plants and their microbial symbionts, as Vittorio Venturi and Christoph Keel report, although the scope could be broadened to include all organisms present.[31] Plants may send signals to the soil microbiome to recruit beneficial species or to activate beneficial traits. In turn, microorganisms may signal to the host plant to activate defenses, influence metabolism and development, or induce the stress response.

Plants are also known to communicate with their neighbors through entangled root networks. Trees, for instance, can feed their offspring through their roots and keep stumps alive. They can also trade information about threats like drought or diseases. Some forest experts have taken to the expression "the wood-wide web" to characterize this hidden information exchange. Just how it works remains to be explored in detail.

Communication may be the most important information-processing function in plants, but it is by no means the only one. Evidence is accumulating to suggest that plants can remember certain kinds of events and learn to ignore them even if those events would normally trigger stress responses. Stefano Mancuso from the University of Florence, Italy, for instance, has trained mimosa (*Mimosa pudica*) plants to tolerate certain kinds of shocks, like being dropped on the floor, without activating their widely known leaf-folding response. After a series of 60 drops, the plants accepted this condition as normal, while still retaining their sensitivity to other unexpected events, such as being touched or shaken. Monica Gagliano's group at the University of Western Australia, Crawley campus, worked with Mancuso to analyze the plant's memory using approaches normally reserved for animal behavior. They found that learning and memory improve in energetically costly environments where these skills matter more.[32]

In 2016, Rainer Hedrich from the University of Würzburg, Germany, and colleagues reported that the Venus flytrap, *Dionaea muscipula*, can count up to five mechanical stimuli received via the sensory hairs inside its trap.[33] One stimulus might come from a non-target source, such as a fallen leaf. Further hits indicate the presence of a moving insect, and after each signal, the plant steps up the response, first closing the trap and then releasing its gastric enzymes to digest the prey. By this stepwise response to the signals received, the plant ensures that it doesn't waste energy on false alarms.

If plants can apparently communicate, remember, and count, can we consider these abilities as cognitive processes even in the absence of a brain? The concept certainly resonates with the zeitgeist—

witness the book by the German forest ranger Peter Wohlleben, *The Hidden Life of Trees*. Wohlleben popularizes what plant science has established so far by unashamedly using anthropomorphisms to describe the cognitive and social interactions of plants. Thus he talks of brood care, friendship, and social networks among trees, which appears to resonate with the forest-loving readership.

The risk of such language is that it may revive controversies over highly esoteric claims made in the past regarding plant consciousness, often based on a mixture of scientific insights and pure fancy. The findings made recently with the most advanced techniques that twenty-first-century science can command do suggest, however, that the cognitive abilities of plants deserve to be taken seriously and studied further. They are not just fantasy.

CHAPTER 2

Fantastic Animals

Life as we know it today has a clear division of labor and ecological roles among multicellular species. Plants, helped by fungi and microbes, as we have seen in chapter 1, are primary producers; they typically absorb carbon dioxide from the air (or water) and use it to build biomolecules, including proteins and carbohydrates. Animals, by contrast, digest these molecules made by plants, gain energy from burning them, and release carbon dioxide into the air. They also excrete nutrients, such as ammonia and phosphates, that can in turn help plants grow.

This nicely symmetrical double cycle of life with plants and animals complementing each other, like yin and yang, has kept evolution going and enabled the diversification of a fascinating variety of species. Among animals, the competition for plant-made food resources and the development of predation and the food web, with multiple layers of predators and prey, have led to the evolution of an impressive functional diversity. Fish learned to walk, and dinosaurs learned to fly, essentially to avoid predators.

Being animals and mammals ourselves, humans may be biased here, but we never stop marveling at the diversity produced by the evolutionary hunger games. This chapter rounds up some of the most fascinating examples of animal diversity. Beyond just admiring the diverse shapes, sizes, and functions, we should bear in mind that all of this is driven by ecology, by the need to eat and to avoid being eaten.

Let's begin with the oldest and simplest animals. Although they are often sidelined as the most "primitive" of multicellular animals, sponges (Porifera) are a key witness to the early evolution of complex life. Filter feeding seawater, they are also important for marine ecology and could serve as bioindicators of ocean health and pollution. Their unique abilities in biomineralization, including the formation of glass spicules at ambient temperature, can inspire modern material science.

Magical mysteries of marine sponges

The Aegean island of Kalymnos has been the home of professional divers for millennia. Due to a shortage of agricultural land on the volcanic ground, residents have always depended on trade, and they most successfully traded what they found on the seafloor around their island: sponges.

The use of natural sponges for cleaning and other purposes is recorded in the literature of Mediterranean antiquity going back to Homer. As ancient Greek and Latin imported the word *spongium* from an unidentified but non-Indo-European language, the whole idea of exploiting these animals may have been a cultural import as well.

The Mediterranean tradition relied on the genera *Spongia* and *Hippospongia*, which have the advantage of being softer than other sponges, as they don't have a mineralized skeleton. In the nineteenth century, additional species abundantly found in the Caribbean also became popular. Natural sponges were widely used until the middle of the twentieth century, when industrial production from synthetic materials came to dominate the market, leaving the divers of Kalymnos to work mainly for the tourist business.

Today, marine sponges are of interest for many other reasons: as representatives of one of the first forms of multicellular life to evolve, as bioindicators of environment health, as sources for bioactives (substances with interesting pharmacological or other activities), and as an inspiration for the enzymatically guided processing of materials, including glass.

Sponges are animals without any differentiated tissues or organs. Like a very early embryo, they only have an inner and an outer layer of cells, and quite a lot of pores to filter seawater. Anything they catch from the water is digested by the cells directly.

It appears plausible to assume this to be the deepest branch of animal evolution, representing the status before differentiated tissues and organs evolved. However, as the early stages of the evolution of animals are poorly represented in the fossil record, there have been debates over the origins of sponges in relation to other animal lineages.

Specifically, some analyses of molecular phylogeny appeared to suggest that comb jellies (Ctenophora), although more differentiated anatomically, might represent an earlier branching. Gert Wörheide's group at LMU Munich, Germany, came to the defense of sponges. In a paper published 2017, they established that inadequate data processing had produced artifactual results suggesting comb jellies to be older than they are.[1]

Further evidence for the early evolutionary presence of sponges came from typical metabolites produced by these animals and discovered as "molecular fossils" in rocks and oils from the Neoproterozoic (660–635 million years ago), more than 100 million years before the Cambrian explosion of animal diversity, including sponges, appearing in the fossil record.

Gordon Love's group at University of California, Riverside, had originally discovered the presence of a steroid biomarker called 24-isopropylcholestane (24-ipc) in marine sediments in south Oman and interpreted this as evidence of early sponges, linking them to today's demosponges, which represent the largest and most diverse class in the phylum of sponges. However, it has been noted that certain algae are also able to produce this molecule.

In 2018, the group discovered a new biomarker, the steroid compound 26-methylstigmastane (26-mes), which occurs together with 24-ipc in Neoproterozoic rocks and has a unique structure that is currently only known to be synthesized by demosponges.[2] Even its metabolic precursors are specific to this class. The authors concluded that these findings "strongly suggest that demosponges, and hence multicellular animals, were prominent in some late Neoproterozoic marine environments at least extending back to the Cryogenian period."

The earliest anatomical fossils assigned to sponges (but also disputed by some) are the Archaeocyatha from the Early Cambrian (around 530 million years ago). They disappeared by the Middle to Late Cambrian and were replaced by rapidly diversifying groups we can recognize as demosponges, which appear in the Burgess Shale, for instance (508 million years ago).

Even though sponges went their separate ways before other animals started evolving organs and nervous systems, they have had just as much time to evolve their own complexities, including elaborate mineral skeletons. Some of the more intriguing examples of their complexity were discovered in recent years in deeper waters, such as the harp sponge (*Chondrocladia lyra*). This species, discovered in 2012, has hooks on its exposed glass skeleton, which serve to trap small crustaceans. Once a prey is caught, the sponge releases a membrane with enzymes to digest it and resorb the nutrients.

Another new sponge species described more recently is the demosponge *Desmacella hyalina*, which thrives in deep waters off the coast of British Columbia, Canada, where it settles on the surface of reefs formed by glass sponges. After Canadian authorities designated the area a marine protected area in 2017 to guard these unique reefs stretching across hundreds of kilometers, Sally Leys's group at the University of Alberta, Canada, set out to characterize the sponges living on top of the reefs using a robotic sampling device.[3] Although one species of *Desmacella* had already been described, the researchers noted that these demosponges came in different color types. Genome analyses revealed that the white color type is a new species, *Desmacella hyalina*. While the ecological importance of the glass reef for crustaceans and fish has been studied in detail, the interaction between the glass sponges forming the reef and the demosponges that live on top of it as epibionts remains to be elucidated.

As sponges could, in principle, live at any depth and much of the deep sea remains unexplored, there may well be many species waiting to be discovered.

While the sponges that divers used to harvest for sale are soft throughout, most species have a hard skeleton made either of calcium minerals or silica, like glass. The latter material, often hidden but sometimes visible, like in the harp sponge and in the hexactinellids (glass sponges), has particularly fascinated marine biologists and material scientists.

Humans have been producing glass since antiquity, always using high temperatures, typically above 1,400°C. So how can the most primitive of all multicellular animals produce delicate glass structures without a heat source?

In 1998, the protein silicatein was discovered in sponges and confirmed as a catalyst of silica deposition in vitro. It is located in the core of glass spicules produced by demosponges like *Tethya aurantium*, and findings from Igor Zlotnikov's group at the Technical University of Dresden, Germany, suggest that fibers formed by this protein are the scaffolding that determines the growth direction and branching of the glass needles.[4]

Structural and functional analyses of silicatein were hindered by its failure to crystallize in the laboratory. In earlier work at the Max Planck Institute of Colloids and Interfaces in Potsdam, Germany, however, Zlotnikov had shown that the protein naturally occurs in a crystalline state in the core of the sponge glass needles. After his move to Dresden, Zlotnikov initiated an effort to solve the crystal structure using the natural material. Stefan Görlich attempted to use the natural needles as one would use a glass tube with a protein crystal in normal X-ray crystallography. This proved impossible, and the researchers had to resort to an alternative technique of combining data from many crystals, known as serial crystallography.

While the structure was largely in agreement with the predictions based on related proteins, it turned out that the presumed active center of the enzyme was turned off in the needles. This finding appears plausible as the enzymatic function is no longer needed once the protein fiber is surrounded by glass, but it also means that the molecular mechanism of silica deposition cannot be deduced from the crystal structure. Thus, further challenges lie ahead to re-create the process of a sponge's glass factory in the laboratory and decode its secrets. Beyond biology, understanding this process could also help material scientists develop new ways of depositing microscopic structures of glass and other materials in the context of nanotechnology.

As sponges actively pump thousands of liters of water per day for filter feeding, they are more exposed to their environment than most animals. This led Gert Wörheide's group to the hypothesis that sponges may be suitable bioindicators for microparticle pollution.

Many species of sponges naturally accumulate microparticles, such as grains of sand, which appear to serve a role in strengthening their structures. Wörheide's group collected tissue samples from 15 individuals of five abundant sponge species growing in shallow waters in Indonesia and known to naturally incorporate foreign microparticles. Removal of a small patch of tissue with a knife does not harm the animal as it can easily regenerate.

The researchers conducted histological analyses on these samples and used microscopy and Raman spectroscopy to identify any microparticles absorbed in the tissue.[5] They found a wide range of particles, including many that were clearly of anthropogenic origin, such as particulate cotton, titanium dioxide, plastic, and blue pigments. While particle size determines retention and location within the sponges, none of the species investigated appeared to have any preference for the accumulation of specific types of materials.

The authors conclude that sponges are suitable as bioindicators for particle pollution, as they are indiscriminate sediment traps reflecting the distribution of particles in the water surrounding them. Their widespread occurrence and exposure to water conditions has also led to suggestions that they could be used as indicators of water quality and chemical pollutants. More surprising, perhaps, is the recent finding that they retain environmental DNA, enabling researchers to monitor biodiversity by proxy.

As Stefano Mariani from the University of Salford in Manchester, United Kingdom, and colleagues have reported, the use of PCR (polymerase chain reaction) with vertebrate-specific primer sequences makes it possible to isolate the DNA of a variety of marine species from sponge tissues.[6] While the majority of the 31 species identified by their DNA barcodes were fishes, the researchers also detected sequences from Weddell seals (*Leptonychotes weddellii*) and chinstrap penguins (*Pygoscelis antarcticus*).

Mariani suggested that sponges could be used as natural samplers in biodiversity studies, instead of using intrusive technical appliances considered for the sampling of environmental DNA, such as robotic submarines.

If Mariani and his co-workers hadn't selectively amplified vertebrate DNA, they would have probably been swamped by microbial genomes, because sponges are now regarded as holobionts that host a range of bacteria, archaea, and viruses—much like the plants we discussed in chapter 1.

The groups of Ute Hentschel at Kiel University and Peter Schupp at the University of Oldenburg, both in Germany, with colleagues including Wörheide, set out to broaden the range of microbiome studies by including those of sponges from deeper waters and colder climates, where archaea play a bigger part as symbionts than has previously been appreciated.[7] The researchers constructed new gene libraries for microbiota of demosponge and hexactinellids sampled from deep water locations in the South Pacific and compared them to literature data of the well-documented shallow-water demosponges. To exclude the possibility of amplifying environmental DNA, such as the vertebrate genomes detected in the study mentioned above, the researchers also applied quantitative PCR.

They found that the bacterial communities in deepwater demosponges generally resemble those found in shallow water, except for the absence of cyanobacteria in the deep water, which is understandable given that cyanobacteria specialize in photosynthesis. A sponge-specific group, the Poribacteria, also seemed to be underrepresented in deep-sea sponges.

Analyzing seven species from the hexactinellids, for which no previous microbiome composition was reported, the researchers found mainly Gammaproteobacteria, with fewer other species than usually found in demosponges. Only two species showed a diverse microbiome with a similar range to that of demosponges.

Both groups were found to host archaea from the family Nitrosopumilaceae (phylum Thaumarchaeota). The quantitative analyses suggest that these are much more important in deep sea than in shallow-water sponges. The researchers hypothesize that thaumarchaeotal ammonia oxidation may be important to the sponges, but the precise role of archaea within the sponge holobiont remains to be elucidated.

Shallow-water sponges are useful to researchers, as they can be grown in aquariums and subjected to experimental changes of conditions such as temperature. In a recent study using the coral reef sponge *Lendenfeldia chondrodes*, Sergio Vargas and colleagues at LMU Munich assessed how the sponge's microbiome responds to rising water temperatures.[8] They found evidence that the microbiome may be resilient toward expected environmental change in the marine habitat.

Some species of soft sponges might have been driven to extinction by human exploitation had their use not been replaced by synthetic materials just in time. Similarly, there are now concerns for deepwater species as human activities, such as mining, expand into deeper waters and threaten to industrialize the oceans.

Swee-Cheng Lim from the University of Singapore and colleagues described a new species assigned to the family Stelligeridae and named *Plenaster craigi*. It grows abundantly on the seafloor of the Pacific in the Clarion-Clipperton Zone, an area identified as a target for mining of polymetallic nodules, grapefruit-sized metal balls containing valuable resources such as manganese, nickel, copper, and cobalt.

Surveys found that sponges like *Plenaster craigi* often encrust these nodules. But even if they are not directly removed with the harvested material, sponges will be at risk from any mining activities on the seafloor. These sponges, living at depths of 4,000 meters (13,000 ft.) or more, depend on very small amounts of nutrients raining down, requiring them to filter even larger amounts of water than the shallow-water species, which makes them vulnerable to the sand and mud stirred up by mining operations.

Other industrial-scale activities, like trawling, are known to disrupt seafloor communities including sponges. The age-old tradition of not even acknowledging these species as living animals, exemplified by sponge divers collecting them as useful devices, may endanger their survival. Although some, like those on the Canadian glass reefs, are benefiting from marine protected areas, there remains a risk that some of this intriguing animal life will be destroyed before we even manage to get to know and understand it.

Friends, foes, and followers of fishes

Larger fish species are often seen accompanied by smaller species—which may be anything from helpful cleaners, via unobtrusive fellow travelers, through to flesh-eating parasites. Some of these ecological relations have been studied in detail, but many, especially the more dynamic ones, remain poorly understood. These connections may be vital for our understanding of the fate of marine species in a changing oceanic environment.

Manta rays, with their vast wingspan of up to seven meters (23 ft.), are among the most charismatic fish species one can observe in the warmer parts of the oceans and thus a key attraction for scuba-diving destinations such as the Maldives or Hawaii. Of the two species now merged into the genus *Mobula* (together with the devil rays), the smaller, more coastal reef manta ray (*Mobula alfredi*) can readily be observed near coral reefs, while the giant ocean manta ray (*Mobula birostris*) is slightly more elusive.

Both species had been listed as vulnerable on the IUCN (International Union for Conservation of Nature) Red List of Threatened Species, but *M. birostris* has now been moved to endangered.

Countries such as Indonesia have realized that the long-lived animals are much more economically valuable alive as a tourist attraction than dead on the fish market and have protected them accordingly. However, a growing demand for the cartilage of the manta ray's gill plates is still fueling an unsustainable level of killing, which often happens either illegally or in international waters.

The Manta Trust, an international charity based in the United Kingdom, is working to improve the protection and appreciation of manta rays. Its experts can even identify thousands of individual manta rays. While the rays are mostly black from above (the dorsal side), their ventral side is typically white with a pattern of black spots that is unique in each individual.

Working with the dive and snorkel tourism industry in the Maldives, the Manta Trust has established an image database identifying individual resident reef manta rays by their spots; the database had passed 5,000 entries in December 2020. As with the starlike patterns of whale sharks and the whisker spots of lions, the rays' spots are like fingerprints that identify individuals, thus enabling researchers to better understand the life cycle of the animals, including their social network within their own species and with other species.

Manta rays are often observed aggregating in groups of several dozen individuals, and they are often seen accompanied by smaller fish species, including some that attach themselves to the rays. These fellow travelers obviously use the large rays for their own benefit, but are they parasites, mutualists, or commensalists (i.e., do they harm, benefit, or have no effect on the rays)? As it is difficult to monitor the contacts of a fast-moving species, surprisingly little is known about the interspecies connection of manta rays.

Aimee Nicholson-Jack from the Manta Trust and colleagues expanded the scope of knowledge on the manta ray "hitchhikers" with what the authors say is only the second systematic analysis of such species to be published in a peer-reviewed format.[9] Using the Maldives database and photographic evidence accumulated in three decades worth of citizen science engagement, the researchers evaluated the associations observed for 4,901 *M. alfredi* individuals identified in 353 sites.

They found 12 species associated with the reef manta rays, including, for the first time, species that don't belong to the well-known family Echeneidae (remoras). The most observed companion was the sharksucker remora (*Echeneis naucrates*), which was observed in 10% of the sightings. It was most likely to be present when the rays were visiting cleaning stations, which are nearshore locations where populations of cleaner fish feed on crustacean parasites present on the rays. Near-term pregnant female rays were more likely to be seen with *E. naucrates*, but this may be causally related with the fact that they spend more time at cleaning stations. Interestingly, juvenile remoras tended to be found with juvenile rays.

In addition to filter feeding near the surface, manta rays dive to depths below 200 meters (660 ft.) to hunt. It is thought that *E. naucrates* is unable to survive at that depth, so the movement pattern of the ray explains why the hitchhikers aren't always with them. A tagging study of *M. alfredi* diving behavior found that all tagged individuals went to depths beyond 300 meters (980 ft.), with a maximum recorded depth of 672 meters (2,205 ft.).[10]

By contrast, the most common companion of the giant oceanic manta ray, the giant remora (*Remora remora*), seems to stay with the ray for longer and often attaches itself. Based on records of 663 identified *M. birostris* individuals, *R. remora* was present in just over half the sightings.

While the remoras provide some benefits to the manta ray host, including removal of crustacean parasites, the habit of attaching to the ray can also come with a fitness cost. The researchers observed examples of scars due to remora attachment and of remoras attached in unsuitable places, such as inside the gills or the anus.

A third remora species, the white suckerfish (*Remora albescens*), was only rarely seen on the images analyzed, but it has been discovered within the mouths of rays examined more closely, which leads the authors to suggest that these remoras are more commonly associated with rays than the observation data suggest.

New hitchhiker species from outside the echeneid family identified for the first time included the rainbow runner (*Elagatis bipinnulata*), cobia (*Rachycentron canadum*), red snapper (*Lutjanus bohar*), Chinese trumpetfish (*Aulostomus chinensis*), and black (*Caranx lugubris*), bluefin (*C. melampygus*), and giant (*C. ignobilis*) trevallies. These species are generally more loosely associated with the rays and not thought to harm them, so their relation would be classified as commensalism. They benefit from the presence of the larger species in terms of shelter from predators (shown for juvenile golden trevally) and availability of food leftovers and may also save energy due to drag reduction. Some, including adult trevally, snapper, and trumpetfish, use the body of the manta ray to get closer to their prey, and they launch their predation attack from this shelter.

Considering that the habitat of manta rays is already changing rapidly due to climate change and other anthropogenic factors, the authors call for a more comprehensive assessment of the ecological network around these charismatic and vulnerable species. The study of the Maldives population, backed by the unique tool of the image database identifying individual animals, is laying a foundation for further research.

One other recent study of manta rays and their companions focused on observations of *Mobula birostris* and its symbiont *Remora remora* in a marine protected area near the Pacific coast of

Mexico. Edgar Becerril-García from the Instituto Politécnico Nacional, La Paz, Mexico, and colleagues analyzed 271 images taken
by researchers and amateur divers and documented a mean of 1.6
remoras per manta ray.[11]

Among the hitchhiking species, the remoras have attracted attention for the efficiency of the reversible suction mechanism they
use to attach to their hosts, which apart from rays also include
other marine megafauna from sharks to whales (animals with
more than 44 kg, or 97 lb., body weight are considered megafauna).
The attachment survives remarkable drag forces on fast-swimming
animals but can be fastened or released in a fraction of a second.
They achieve this with an adhesive disc that evolved from dorsal
fin elements and consists of a series of parallel lamellae surrounded by a fleshy lip. When the lip makes contact, the lamellae
rotate to expand the enclosed space and thus reduce the pressure.

Brooke Flammang's group at New Jersey Institute of Technology in Newark has studied the suction apparatus for clues to remora behavior, especially when, where, and why they attach or let
go. Investigating the anatomy of the lip in *E. naucrates*, the researchers discovered densely innervated structures that they propose to be push-rod mechanoreceptors that allow the remoras to
sense when they have made contact with a host and quickly trigger
attachment.[12] Once attached, the mechanism would enable the
remoras to sense shear forces and ensure they remain fixed.

While entirely plausible in terms of what the remoras need to
hitch a ride, this was a sensational discovery because it is unprecedented among fishes. The only other comparable mechanism the
researchers could find in the literature is in monotremes (platypus
and echidnae).

Regarding the question where remoras attach and how they be-
have on their host surface, Flammang had a lucky break when she
saw a conference talk from whale researcher Jeremy Goldbogen of

Stanford University, who had tagged blue whales with video cameras and "inadvertently gotten hundreds of hours of remora footage."[13] After seeing the movies of remoras skating on the surface of the whales, Flammang set out to analyze their behavior. Until then, all the information available was based on photos and short-term observations.

Working with experts on fluid dynamics and the Barcelona Supercomputing Center in Spain, Flammang and colleagues obtained models of the flow conditions around the whale.[14] They identified various areas where the anatomy of the whale reduces the drag forces and makes it easier for the remoras to remain attached, such as behind the blowhole and behind the fins. In these areas, the drag force on a whale swimming at 1.5 meters (5 ft.) per second is reduced by up to 80%. Sure enough, the videos show the remoras preferentially settling in these areas, thus saving energy. Calculations showed, however, that they could maintain attachment anywhere on the whale, even on the highly agile tail fluke.

While the remoras had their favorite spots, they didn't always stay in one place. Footage shows them gliding across the surface of their host, to get from one sweet spot to another, or to meet with fellow passengers. Knowing the best spot to attach to animals like whales, rays, and shark will prove helpful to conservation research, as it can facilitate tagging the animals. With applications like tagging in mind, Flammang's group has developed biomimetic suction devices based on the remora studies.[15]

One ecological connection between fish species that has been studied extensively is that between cleaner fish, such as the cleaner wrasses (genus *Labroides*), and their clients, which is a mutualistic relationship, as the cleaners feed by ridding their clients of crustacean parasites. Manta rays and other large fish species visit cleaning stations, environments where cleaner fish are abundant, so this interaction is readily observed at specific locations.

Many other interactions remain to be explored. Some involve parasitic species hiding inside a fish's body cavities, such as the tongue biters, crustaceans of the isopod family Cymothoidae that attach themselves to the host's tongue and often end up eating it and taking its place. These can often remain hidden, as demonstrated by the surprising discovery of a rare species found in a museum specimen almost antipodal to the place where the only other examples were found. The cymothoid *Elthusa splendida* was first described in 1981 based on five specimens recovered from a Cuban dogfish, a deep-sea shark, that had been caught in the western South Atlantic, off southern Brazil. Among cymothoids settling in the mouth of their host, it is an unusual species as it attaches to the palate.

No other representatives of this species were reported until November 2020, when Ryota Kawanishi and Shinpei Ohashi from Hokkaido University, Japan, identified one they discovered in a museum specimen of another deep-sea shark, a Japanese spurdog, caught in the East China Sea, very nearly on the opposite side of the planet compared to the original discovery.[16] This surprising discovery appears to suggest that the species is, or at least has been, present around the globe but has managed to remain hidden from the eyes of scientists. The authors suggest a more generalized scrutiny of fish specimens kept in museums, in case other examples are still awaiting discovery.

Other connections between species remain elusive because they are fleeting and dynamic. Thus, large fish species like whale sharks (*Rhincodon typus*) are often observed in the company of numerous smaller species, including the blackfin tuna (*Thunnus atlanticus*), which in a survey conducted in the Gulf of Mexico was the whale shark's most common companion among more than a dozen species identified. A detailed analysis of the precise nature of the interactions of those species with the whale shark remains to be accomplished.

The typical Pleistocene megafauna has so far survived much better in the marine environment than on land, although several whale species had a close encounter with extinction. As the impact of anthropogenic change continues to increase, efforts to save these species will require, among other things, a better understanding of the complex network of interactions that links the charismatic big beasts to the other fish in the oceans.

Save the dragons

Dragons are described in the myths and legends of cultures around the world, based on ideas rooted deep in prehistory. European dragon stories, for instance, can be traced back to the shared Indo-European tradition of stories. East Asia's dragons have similarly deep cultural roots in the prehistory of the continent.

Therefore, the origins of the concept are probably lost forever. Competing theories attribute them variously to discoveries of dinosaur or other megafaunal fossils, misrepresentation of real-life large reptiles, such as crocodiles, that have since become extinct, and to an instinctive human fear of snakes and other reptiles.

While Chinese and other Asian traditions attribute high intelligence to dragons, and China has associated them with the emperors since the Han dynasty (202 BCE–220 CE), the medieval European thinking typically imagines them as monsters that human heroes are called upon to kill. In this view, it was safest to imagine their habitat in faraway, unexplored lands, hence the occasional inclusion of dragons (as well as other dangerous creatures) on old maps.

As modern European explorers became aware of the biodiversity in tropical lands and islands, they also discovered numerous large and unusual reptile species, some of which, like the Komodo dragon (*Varanus komodoensis*), they named after the mythical beasts. Some of these have only been scientifically described in the twentieth century (including the Komodo dragon in 1912), and some are already at risk of extinction. Rather than slaying, many of the real-life dragons now need saving.

Komodo dragons currently live on five Indonesian islands. They occur in areas on the northern coast of the larger island of Flores, as well as on four islands located within the Komodo National Park, namely Komodo, Rinca, Nusa Kode, and Gili Motang.

At the end of the rainy season (March to April), when there will be abundant new foliage and insects for them to eat, Komodo dragons hatch from eggs buried deep underground. The hatchlings are just 40 centimeters (16 in.) long and weigh around 100 grams (3.5 oz.). They make their way into the trees, where they spend their first year feeding on insects and small lizards.

As they grow, their lifestyle adapts accordingly. After a year, they move to mostly feeding on the ground, where they may eat snakes as well as lizards. From when they weigh around 20 kilograms (44 lb.), the Komodo dragons will live mainly on the ground and prey on larger animals too. Whereas females invest much of their energy in reproduction and forgo food while watching over newly laid eggs, males keep growing for many years and can reach a length of three meters (10 ft.) with a body weight of up to 100 kilograms (220 lb.).

Thanks to a toxin and bacteria present in their saliva, the reptiles can ambush and stun virtually any prey, such that adult Komodo dragons will routinely kill deer and pigs, and in some areas even water buffalo. As apex predators of their islands, the dragons are sensitive to population changes of the larger prey species. Thus, on one island, the overhunting of deer by human settlers has caused the Komodo dragon to be wiped out.

While the ecological requirements of the Komodo dragon are already fairly well understood, the sensitivity of the remaining island populations to climate change hasn't been studied until recently, even though it is already clear that island populations in general are more vulnerable to climate change than continental ones.

Alice Jones from the University of Adelaide, Australia, and colleagues have modeled the likely effect of climate change on the remaining populations of Komodo dragons, taking into account the inherent uncertainties of such models.[17] Based solely on direct effects of environmental conditions on the target species (as opposed to indirect effects via prey species), the researchers found that under all climate models, including the most optimistic ones, the populations will be decimated.

The main cause of direct climate effects is the warming of the dragons' natural habitat, which is the border between open coastal land and inland forest. As the reptiles have no temperature regulation, they would have to migrate to remain in their optimal temperature range, and the islands may not be able to provide them with suitable habitat at the right temperature under standard climate-change scenarios.

In the more optimistic scenarios, the effect would just be a population bottleneck, and conservation scientists would have to watch out for the genetic viability of the reduced populations and possibly intervene with transpositions. In the scenarios predicting faster temperature increases, the species may well find itself on the brink of extinction by 2050. The authors therefore call for urgent protection and mitigation measures, predicting that the islands of Komodo and Rinca may offer the best "future-proofed" habitat for these iconic reptiles. In the updated Red List released in September 2021, the IUCN moved the Komodo dragon from vulnerable to endangered.

The genus *Varanus*—which includes around 80 species with a remarkable size distribution, from the tiny, short-tailed pygmy monitor (*Varanus brevicauda*) to the Komodo dragon—is distributed around the tropics and subtropics. Its name goes back to a common Semitic root found in Arabic as *waran, waral,* or *warar,* as well as in Aramaic as *warna.* Interestingly, German scholars of the nineteenth century appear to have misappropriated the Arabic term to form the German *Warner* (one who warns), or *Warneidechse* (a warning lizard). Translated into Latin, this gave us the original name, *Monitor,* for the genus, which survives in the common English names of many species, even though official nomenclature later replaced it with *Varanus.* Some etymological explanations have also retrospectively attributed the name "monitor" to the habit of the lizards of standing upright on their hind legs and tail.

Among the "warners" related to the Komodo dragon are the Australian sand monitors, which have also suffered a second etymological mishap in that they are called goannas, which derives from the (unrelated) iguana. Genome studies on a group of sand monitor species suggest the Komodo dragon may have originated in Australia. Carlos Pavón-Vázquez from the Australian National University in Canberra and colleagues analyzed nuclear and mitochondrial genomes as well as phenotype data and fossils to clarify the relationship between the Komodo dragon and the sand goanna (*Varanus gouldii*), as well as related species colloquially known as sand monitors.[18]

They discovered genetic traces that were incompatible with a simple bifurcating family tree of the *Varanus* genus. Instead, they concluded that Komodo dragons must have lived in northern Australia and hybridized with the common ancestor of the sand monitor species studied. This is also in agreement with the fossil record and suggests that the Indonesian islands are the giant lizard's last refuge rather than its place of origin. Considering their population has become extinct in Australia, this is a further warning to look after the remaining populations and keep Komodo dragons safe.

Another *Varanus* species that is interesting to compare to the Komodo dragon is the Asian water monitor (*Varanus salvator*). Growing almost as big as the Komodo dragon, the water monitor is widely distributed across Southeast Asia and Oceania, apparently coping well with human disturbance and land-use change. Even though it is hunted for its skin and meat in many places, the species is thriving to an extent that it is one of the most observed wildlife species in Malaysia and is listed as least concern by the IUCN. However, the phylogenetic status of the species is still in flux, as several new subspecies were identified recently, and some subspecies have become new species. Thus, if any of the more marginal populations were to be recognized as a separate species, these could conceivably become a conservation concern.

In the Malaysian part of Borneo, in the Kinabatangan floodplain, the water monitor occurs in a highly fragmented natural landscape, as nearly half the land surface has been converted to oil palm plantations. Several studies have found that this human disturbance has not noticeably harmed the water monitors. Although the plantations don't offer much shelter to the lizards, they do provide abundant prey, such as rodents.

Sergio Guerrero-Sanchez from Cardiff University, United Kingdom, and colleagues presented a detailed population study of water monitors in the oil palm plantations and in the remaining fragments of native forest.[19] In contrast to some earlier studies, the research showed no higher population density in the plantations compared to the forest. The authors concluded that there are source-sink dynamics between both types of habitats, with the forests providing important refuges and breeding grounds, while the plantations and the absence of competing predators are helping to keep the water monitors fed.

Their analysis of body sizes points in the same direction. There appears to be no overall difference between animals found in forests or plantations, but the authors found among forest dwellers a slight advantage for those living in smaller fragments with better access to the plantations as food sources.

While the results suggest the water monitors are among the beneficiaries of the human disturbance of the landscape, they also highlight the importance of diversity and connectivity between different landscape elements. If the remaining half of the landscape were to be turned into plantations, the lizards would likely suffer for lack of shelter.

Other Australian lizard species that have been named after the dragons of myth include the bearded dragons, six species of the genus *Pogona*, which are all doing well and are listed by the IUCN as least concern. The central bearded dragon (*Pogona vitticeps*) has recently emerged as a useful model species for a near-magical capability. Although it has a chromosome-based sex determination system, this can be overridden by a separate mechanism, turning genetically male embryos into females.

Sarah Whiteley from the University of Canberra, Australia, and colleagues have analyzed transcriptomes of the separate pathways on which chromosomal and temperature-induced sex determination converge on the same phenotype, the development of ovaries.[20] The results show that the two pathways initially differ both in the timing and in the choice of genes involved. They also corroborate an earlier hypothesis that calcium signaling and redox state combine to play an important role in linking temperature sensing to the causation of sex reversal.

Environmental sex determination is widespread in many fish and reptile species, but the molecular mechanisms of its initiation have remained elusive. The bearded dragon with its parallel pathways now provides a model to elucidate the causation, and the transcriptome study offers several candidate genes that could be targeted in future functional analyses.

If dragons are unique and special creatures rooted in a remote and obscure past, then the tuatara (*Sphenodon punctatus*) may well be their closest relative in the real world. While it looks like a lizard, it is only distantly related to these reptiles. Its lineage diverged from the common ancestor of lizards and snakes (squamates) some 250 million years ago, in the Early Triassic. The tuatara has long puzzled zoologist with a suite of unusual traits, including surprising longevity, low metabolic rate, and temperature-dependent sex determination (producing females below 22°C, and above that, males).

Neil Gemmell from the University of Otago, New Zealand, and colleagues reported the first assembly of the genome sequence of the tuatara in 2020. As the species holds special cultural significance for the local Ngātiwai people, the international team of researchers also worked with representatives of the group, via the Ngātiwai Trust Board.[21]

The genome assembly uncovered a mixture of features previously held to be specific to reptiles or to mammals, confirming the tuatara's unique position in the tree of life. It is the only surviving species of the order Rhynchocephalia, which was once widespread on the supercontinent of Gondwana and ranks as a sister group to the squamates.

The genome also highlights several unique and interesting aspects of the ancient animal's biology, including an unusual emphasis on color vision (given that it is a night hunter), a slow rate of evolution, and its remarkable longevity and disease resistance.

In terms of conservation concerns, the genome bears traces of recent highs and lows in the effective population size, and the northern population in particular has suffered from recent bottlenecks. While the current abundance of the animals earns it an IUCN least concern listing, there are aspects of vulnerability, including the temperature-dependent sex determination (which could obviously be derailed by climate change) and the dangers of habitat loss and invasive species. Even at a low threat level, the evolutionary distinctiveness of this species means we should look after it.

How sloths got their sloth

Considering that they are named after one of the seven deadly sins, sloths are a remarkably popular group of mammals. Perhaps this is because their life in the slow lane reflects a sin that many humans are ready to admit to and might even aspire to.

Nevertheless, scientists studying these unusual creatures have a lot of work to do and no time to be lazy. The upside-down life of sloths typically suspended from tree branches reflects a wider range of unique traits that set them apart from other mammals. Their ecology and evolutionary history are also remarkably complex.

The six extant species grouped in two genera are the survivors of a much larger suborder (Folivora) that included ground-living and even aquatic relatives. Some years ago, cladistic studies based on anatomical traits as revealed by comparative anatomy of museum specimens and the fossil record had reached a consensus on phylogenetic relationships within this group. But now two studies based on different and independent kinds of molecular analyses have shown a very different pattern of family connections.

The numerous extinct relatives of the modern-day sloths comprising Folivora cover a broad range of ecological roles, from elephant-sized giant sloths to smaller forms, most of them ground-living and some even semi-aquatic. Many of these species were widespread across the Americas and only became extinct after the arrival of humans at the end of the Pleistocene. Therefore, fossil and even tissue remains of these animals are abundant and often seen in natural history museums.

Their closest relatives are the anteaters (Vermilingua) within the order Pilosa, which together with the armadillos (Cingulata) complete the superorder Xenarthra. Morphological analyses of the numerous fossils had suggested a family tree that placed the three-fingered sloths (four species of the genus *Bradypus*) on an early branch diverging from the rest of the sloths, while the two-fingered sloths (two species in the genus *Choloepus*) were considered part of the family Megalonychidae, including giant ground sloths (e.g., *Megalonyx jeffersonii*) and smaller Caribbean endemic sloths (e.g., *Acratocnus* and *Parocnus*). *Bradypus* and *Choloepus* are widely known as three- and two-toed sloths, respectively, even though the distinguishing digits are on their fore limbs, so referring to them as fingers is more accurate.

Preliminary ancient DNA evidence, produced in the early 2000s, appeared irreconcilable with this family tree, although not strong or detailed enough to overthrow it. By 2019, two independent studies, one using mitochondrial genomes and the other collagen amino acid sequences, consistently and convincingly suggested an entirely different story of sloth evolution.

The first complete mitochondrial genome obtained from an extinct sloth species was from a 13,000-year-old bone fragment of *Mylodon darwinii*, found exceptionally well-preserved in the Cueva del Milodón (Mylodon's Cave) in Chile (named after the numerous fossils found there). The research groups led by Hendrik Poinar at McMaster University in Canada and Frédéric Delsuc at Montpellier University in France reported that the mitochondrial genome and preliminary data from the nuclear genome conflicted with the established morphological classification.[22] Other initial studies of specific extinct sloth species were also difficult to reconcile with the morphological version of their evolutionary history.

To clarify the issue, Poinar's and Delsuc's teams analyzed the mitochondrial genomes of 10 extinct sloths, representing six species, in comparison with those of the surviving ones, and came up with a dramatically different family tree, where phylogenetic relationships are drastically rearranged.[23]

While they were analyzing their findings, the researchers were aware that a separate effort focusing on amino acid sequences of the protein collagen type I, which is more abundant in bones and often survives longer and in more challenging circumstances than the DNA, was underway. The investigation led by Ross MacPhee at the American Museum of Natural History in New York came to very similar conclusions.[24] Considering the unexpected nature of their findings, both teams agreed to publish their papers simultaneously so that each could point to the other as an entirely independent confirmation.

What is still true based on the new molecular analyses is that the two genera of living tree sloths are only very distantly related and developed their unique suspensory lifestyle as an adaptation to arboreality independently. Their kinship links to extinct ground-living species will need to be revised, however. Poinar and Delsuc suggested a new taxonomy with eight families and three super-families, and they call on paleontologists to test their morphological data against this new molecular framework.

These studies also cast new light on the likely history of species distribution in South America and the Greater Antilles. The molecular evidence from both studies suggests that the extinct sloths of the Caribbean, the megalocnids, branched out early, more than 30 million years ago. They may have reached the West Indies across a temporary land bridge known as GAARlandia, where GAAR is an acronym for Greater Antilles–Aves Ridge, which is postulated to have existed between 35 million and 33 million years ago. Phylogenetic trees of other Caribbean mammals have failed to match dispersal times with this hypothesis, so the sloths provide the first such study to confirm this biogeographic land connection.

Further molecular analyses may help to clarify these connections and to elucidate the origins of some of the peculiar morphological, ecological, and physiological adaptations that make the
surviving species of tree sloths so unique.

One of the main reasons to study sloth evolution is in the animals' unique tree-based ecology, which both surviving genera appear to have arrived at by convergent evolution. Living on trees and
eating mainly leaves is a lifestyle found in very few mammals, so
studies of sloths can give key insights into the limitations of this
ecological niche.

The poor nutrient quality of a diet based mainly on tree leaves
limits the size range of animals adapting to it. In both genera of
sloth it has led to a range of energy-saving measures, including
slow movements, slow metabolic rate, variable body temperature,
and a specially adapted digestive system.

Jonathan Pauli and colleagues at the University of Wisconsin–
Madison studied these adaptations to the low-energy life in the
trees in detail, showing that the three-fingered sloths, which are
more narrowly specialized in their diet, are more radical energy
savers than two-fingered sloths.[25]

Measuring the field metabolic rate for both species in the wild,
the researchers found that three-fingered sloths had the lowest
value observed for any mammal. They also had more variable body
temperature than two-fingered sloths and moved even less. All
findings suggest that the three-fingered sloths represent the more
extreme example of the sloth lifestyle.

One might worry that species that are so extremely specialized
that they only eat the leaves of one or two species of tree would be
most vulnerable to environmental change. The situation is somewhat more complex for the three-fingered sloths in Costa Rica
(*Bradypus variegatus*), however, as their favorite tree, the guarumo
tree (*Cecropia obtusifolia*), is a fast-growing plant that is quick to

colonize disturbed environments and is also planted in cocoa plantations to provide shade. For sloths, the tree has many advantages, including the fact that it grows new leaves all year round, enabling sloths to feed on the young leaves, which are better digestible.

To establish just how strongly the welfare of the sloths is linked to this specific tree species, Mario Garcés-Restrepo with Jonathan Pauli and Zachariah Peery, all at the University of Wisconsin, have conducted a long-term study on a field site in Costa Rica with a variety of different habitats to establish the fitness of *B. variegatus* as a function of the density of guarumo trees in their range.[26]

The researchers found that both the survival of adult sloths and their reproductive success are strongly linked to the density of the guarumo trees. By contrast, juveniles are not that closely attached to guarumo trees, and their survival does not correlate with the presence of these trees. Garcés-Restrepo and colleagues conclude that the young sloths are likely having to make a trade-off between the better security from predation in other, more densely covered trees against the preferred food available in the guarumo trees.

In a separate study using the same field area, the same authors studied the dispersal of the young animals as they become independent, both for three-fingered (*B. variegatus*) and for two-fingered sloths (*Choloepus hoffmanni*). The study showed that juveniles of both species preferred tropical forest habitat and avoided pastures.[27] The less specialized two-fingered sloths were happy to accept shaded cacao plantations as habitat.

Analyzing the survival rates of young sloths, the researchers found that the animals were most at risk immediately after they became independent of their mothers. Survival chances were lower for the three-fingered than for the two-fingered sloths, but in both cases good enough to support a stable population size.

Sloths are in turn ecologically important for a whole range of other species. Some of the complexity came to light when Pauli's group investigated why the lazier and more specialized three-fingered sloth descends from its tree once a week to defecate and bury its feces underneath its own tree, while two-fingered sloths have been observed just letting their business drop from where they are. For the three-fingered sloth, the toilet trip carries both a significant energy cost and an additional risk of becoming prey to jaguars. Thus, the researchers argued, there must be a major fitness benefit to balance the disadvantages caused by this behavior.

Pauli and colleagues found a plausible explanation in the ecology of the moths and algae that colonize the fur of the three-fingered sloth to a much greater extent than that of the two-fingered ones.[28] The researchers found that the life cycle of the resident moths depends on the sloth toilet behavior—they lay their eggs in the dung the sloth leaves on the ground, and the fledgling moths fly up to find a sloth in the tree above. Enriching the sloth fur with nitrogen, the moths enable the growth of algae. These may help with the camouflage, but the 2014 study by Pauli and colleagues showed that three-fingered sloths consume some of the algae in their fur and obtain valuable nutrients from them, supplementing their limited diet of guarumo leaves. Thus, by virtue of being too lazy to clean their fur, the sloths manage to support an entire ecosystem that ultimately helps their own survival. And the jaguars win too.

Although their narrow specialization and low energy for movement might evoke fears for their survival, only one of the six species is threatened, namely the critically endangered pygmy three-fingered sloth, *Bradypus pygmaeus*. Found exclusively in mangrove forests on the island Escudo de Veraguas off the Caribbean coast of Panama, this species is threatened by the loss of its mangrove habitat and appears in the EDGE (Evolutionarily Distinct and Globally Endangered) list of the most distinct and endangered mammals (http://www.edgeofexistence.org/; more about the EDGE lists in chapter 5).

The IUCN Red List assesses the maned three-fingered sloth (*Bradypus torquatus*) as vulnerable, and the remaining four species are of least concern. No tree sloth species is known to have become extinct. Their ground-living relatives were less lucky.

Rich treasures of relatively recent remains and fossils of ground sloths across the Americas bear witness of a range of species that disappeared at the beginning of the Holocene, soon after human hunters arrived. While their extinctions may in part be related to the climate change marking the end of the Pleistocene, the remains found point to severe hunting pressure that is likely to have contributed to their doom.

Until humans arrived in the Americas, there were numerous species of ground sloths, traditionally arranged in three families, although according to the molecular evidence there were at least eight.

Megatherium species, for instance, evolved during the Pleistocene to reach ever larger sizes, culminating in the late Pleistocene *Megatherium americanum*, which was heavier than today's African elephants. As museum exhibits, the skeletons and reconstructions predate the discovery of dinosaurs, which explains why Georges Cuvier (1769–1832) coined the name *Megatherium*, which simply means "big beast." A rib bone from a 19,000-year-old *M. americanum* from Argentina was one of the fossils that yielded mitochondrial DNA for the genome study by Delsuc and colleagues discussed above.

Giant ground sloths like *M. americanum* were herbivores, consuming tree leaves among other things, but given their size and ability to stand on their hind legs, they had no need to climb trees to reach the leaves. With their size and powerful claws, they would have had no predators to fear—until human hunters arrived.

The only aquatic genus of sloth, *Thalassocnus*, died out three million years ago, long before humans arrived. It evolved several marine adaptations over time and fed on seagrasses off the Pacific coast of South America. The change to a marine diet may have been triggered by the desertification of its terrestrial habitat. Five species were described with various degrees of aquatic adaptation, but they may all belong to the same lineage and represent different stages in its evolution.

Thalassocnus became extinct when South America connected to Central and North America, as the change in ocean currents affected the seagrasses it fed on. At the same time, this shift gave the South American ground sloths the opportunity to expand into Central and North America.

Among the known extinct sloth relatives, the early extinction of *Thalassocnus* remains an outlier. Most of the ground sloths thrived through the Pleistocene until humans arrived—on some Caribbean Islands, they survived to as late as 5,000 years ago.

When hunters with spears started decimating the American megafauna, the sloth lifestyle of dangling from the trees became advantageous in unexpected ways. Tree sloths turned out to be unattractive targets for hunting because, apart from being smaller, their claws may keep them suspended in a tree even after they are killed. A hunter might lose a spear without attaining the anticipated meat ration—hunters must have learned that lesson swiftly.

Thus, at the beginning of the Holocene, the unusual ecological niche found up the trees was the only one where the previously large suborder of sloths could survive. Being lazy can save lives.

Fantastic species and where to find them

In the fictional textbook of "magizoology," *Fantastic Beasts and Where to Find Them*, magizoologist Newt Scamander covers 85 magical species from Acromantula to Yeti, including various kinds of dragons, unicorns, and werewolves. Originally mentioned as a school textbook in the Harry Potter series, the fictional book took on a life of its own when J. K. Rowling published it as a charity stunt in 2001, which eventually led to the release of the eponymous feature film in 2016.

Many of these imaginary creatures are based on ancient myths and legends, often only one step removed from the zoology of the real world. Thus, the tusks that over centuries helped to feed the myths around unicorns came from narwhals, while fossilized remains of various extinct megafauna may have inspired legends of dragons and other magic beasts.

Large parts of the ancient myths can be interpreted as the consequence of incomplete zoological knowledge. When white spots on the map were inscribed "Here be dragons," the gaps in humanity's knowledge were filled in using pure imagination, creating unicorns based on tusks alone.

Back in the real world, as scientists progress to discover and describe more and more of the species that were once hidden in the depths of the oceans or in impenetrable jungles, the sense of awe and wonder inspired by the ancient myths and their modern CGI-enhanced reincarnations could equally well be fed by the exploration of real zoology. The oceans, for instance, are full of fantastic beasts—one has only to figure out where to find them.

Legends were right in that the fantastically straight, spiraled tusks belonged to a single-horned mammalian species. They only got the shape slightly wrong, as the animal behind the impressive spear is not of the elegant equine sort but a rather plump whale.

Living in Arctic waters, the narwhal (*Monodon monoceros*) can work its own kind of magic. It produces ultrasound clicking, which it uses for echolocation. As Jens Koblitz from the Bioacoustic Network in Neuss, Germany, and colleagues established with acoustic measurements in Baffin Bay, West Greenland, the narwhal's sonar is the most strongly focused sound beam produced by any species investigated so far.[29]

Until now, narwhals have been able to live in the Arctic waters relatively undisturbed, following their seasonal migration patterns, typically spending the summer in the High Arctic waters and congregating on winter feeding grounds in Baffin Bay, where deep waters allow them to feed at around 1,500 meters (5,000 ft.) below the surface. The seafloor below the area acts as an acoustic bowl, making it particularly suitable for ultrasound navigation and prey detection.

Considering the dramatic loss of sea ice in the Arctic due to climate change, which is already leading to increased commercial interest in the area for tourism, new shipping routes, and resource extraction, there is a danger that cetaceans like the narwhal may suffer the impact of increased shipping traffic, as the noise pollution from ships may interfere with their acoustic navigation system.

This was part of the motivation of the detailed investigation of Koblitz and colleagues into the current, still undisturbed sonar use of narwhals. It is crucial as a baseline for future studies aiming to detect how the anticipated increase in shipping in Arctic waters and through the Northwest Passage will affect the natural behavior and hunting success of the narwhal. All too soon, such basic understanding could become essential in efforts to save the species—after all, we don't want these very real animals to become the stuff of myth and legends like the unicorns (more about narwhals and their use of sound in chapter 8).

While the dragons in old tales tend to be large and terrifying, with their indomitable fighting spirit and the scientifically unexplained ability to spout fire, real-world seadragons are comparatively small and harmless kinds of fish. Two species of seadragons have long been known to scuba divers exploring coral reefs off the southern and western coasts of Australia, namely the common seadragon (*Phyllopteryx taeniolatus*), and the more elaborately camouflaged leafy seadragon (*Phycodurus eques*), which grow to adult lengths of up to 45 centimeters (18 in.) and 24 centimeters (9 in.), respectively. With their unusual shapes (for a fish) and plantlike appendages, they look like they might have sprung up from the imagination of some writer of fantastic fiction, but they are in fact very real and easy enough to find. Both are listed as near threatened on the IUCN Red List.

In 2015, Greg Rouse and colleagues from the Scripps Institute of Oceanography in San Diego, identified a third related species, the ruby seadragon, *Phyllopteryx dewysea*, which they first described based on four museum specimens. Two years later, they could report the first video observation of the species in its natural habitat, at the Recherche Archipelago off the south coast of Western Australia. Using a small robotic device, the researchers discovered two active individuals of the species at a depth of 54 meters (177 ft.).[30] This is out of range for recreational scuba diving, explaining why the species hadn't been reported before.

Reflecting the less leafy nature of its habitat, the ruby seadragon has no appendages for camouflage but appears to rely on its red hue making it less visible in the poor light of the ocean's twilight zone. Another intriguing feature of the new dragon species is its prehensile tail, which the other two species lack. With it, the animals can anchor themselves against strong currents, which may explain why the ruby seadragon doesn't wash up on beaches as frequently as its relatives. Whether the ruby seadragon kept or reacquired this feature also seen in other species of the Syngnathidae family (pipefishes, seahorses, and seadragons) remains to be established.

The similarly eccentric looking seahorses not only boast the prehensile tail—they also feature a whole range of other morphological traits not usually found in fish, including an elongated snout made of fused jaws with a small terminal mouth, bony plates instead of scales, fins in unusual places, and the famous brood pouch in which the male incubates the spawn. A genome study of the tiger tail seahorse (*Hippocampus comes*) showed that the magic of the seahorses can in part be explained by an unusually rapid rate of evolutionary change.[31]

One of the most intriguing aspects of biodiversity in the oceans is the dramatic display of color not only in tropical fish appreciated by aquarium lovers but also in invertebrates such as the jellyfish and nudibranchs (sea slugs). The vast diversity of species-specific colors and patterns among these species dramatically visualizes the marine biodiversity of which we are only beginning to appreciate a small fraction.

For instance, in a survey of jellyfish found around the coasts of South America, Antonio Carlos Marques from the University of São Paulo, Brazil, compiled 958 morphological types, of which 800 were identified as species. The study included both the phylum of Ctenophora (comb jellies) and the subphylum Medusozoa from the Cnidaria phylum. Medusozoa species have a life cycle that includes a jellyfish-like phase, while other members of Cnidaria, which were not included in the study, don't.[32]

Apart from their colorful and translucent beauty, the jellyfish impressed researchers with a wide range of sizes, from one millimeter (0.04 in.) to one meter (3.3 ft.). While most are harmless to humans, as long as they are not disturbed, some produce powerful toxins that can become fatal. Results of the census are now compiled in an interactive database to which researchers will continue to add new finds.

Similarly, the nudibranchs, named after their peculiar, exposed gills, are expanding in colorful variety each time researchers dive in the oceans to look for more. They are found around the globe in a wide range of habitats, from the intertidal zones to depths beyond 2,000 meters (6,600 ft.). Many species have been described off the coast of California, but in an excursion to the Philippines, Terry Gosliner's group from the California Academy of Sciences could identify over 40 new species, adding further color to the palette.

For many researchers, the very existence—and the current threats to its survival—of biodiversity is sufficient reason to study it. While the terrestrial biosphere is already severely diminished by human impact, most noticeably in the disappearance of mega-fauna in all continents except Africa, the marine environment has been protected by how inaccessible it is. Therefore, it has pre-served an impression of Pleistocene species richness on a scale no longer available on land—but now also threatened by expansion of human activities into the oceans.

If further reasons are needed to investigate the richness of the living world, there are plenty of unexpected benefits to be discov-ered, providing additional justification for curiosity-driven re-search. Thus the genome study of a marine crustacean, *Parhyale hawaiensis*, not only yielded insights into limb regeneration, a "magical" ability fairly widespread among invertebrates, but also uncovered an unexpected biochemical pathway for the degrada-tion of lignocellulose, the indigestible fabric of wood. Biotechno-logical adaptation of this discovery could help turn wood into food or biofuel. The ability to digest lignocellulose, conspicuously ab-sent in mammals like us, could also be described as a special kind of magic.

Wonders of the living world also include electrical wires and biological light bulbs. Deep-sea species have evolved a variety of light organs for different purposes, from the glowing bait of the angler fish to the strong light of the flashlight fish used for hunting. Intriguingly, the smalltooth dragonfish, *Pachystomias microdon*, has three organs producing different colors of light. The blue one appears to be an all-purpose light and can be seen by other species as well. The red light would be invisible to other species and is thus presumed to serve in stealth hunting or in private courtship. The species also carries an orange light, for which scientists have not discovered a use yet, as Liz Langley reported in National Geographic in 2017.[33] Bioluminescence has already found manifold uses in biology and biotechnology, the most prominent example being the green fluorescent protein, originally found in the jellyfish *Aequorea victoria*.

New discoveries from insufficiently explored depths of the oceans could help inspire the next generations of writers to dream up even more fantastic beasts. But all those who want to feed their sense of wonder do not need to turn to fictional species, as our biosphere still has many fantastic species to offer that we should appreciate and protect before they pass from the real world into the realm of legend.

CHAPTER 3

Insects Rule the World

In 2018, researchers launched the Earth BioGenome Project, an ambitious attempt to sequence the genomes of all eukaryotic species known to science, 1.5 million and counting, within 10 years. Adjusting for inflation, the cost is projected to remain just below that of the Human Genome Project—another moonshot for science.

One of the main challenges for the project arises from evolution's widely appreciated fondness for beetles. There are more than 400,000 named species of coleopterans, which are part of the overwhelming count of the even bigger swarm of more than one million species of insects. The genome sequencers will have to spend most of their time and data processing capacity dealing with six-legged life forms.

Although insects represent most of today's animal biodiversity, many of their species are now at risk from land-use change and pesticides. Given their vast number of species and even bigger number of connections to other species, insects form a vital part of the complex ecological network that is life on Earth. It is no wonder that science is still busy finding new connections in their ecology and evolution, including in the ways they co-evolve with plants and other organisms. We also need to improve our understanding of how we can live with them and share the planet peacefully.

Six–legged success stories

Life on Earth, from our anthropocentric perspective, is often described as the evolution toward "higher," more intelligent life forms, with humans as the alleged pinnacle of creation. Alien visitors passing through the Solar System might first spot the green vegetation and measure its oxygen output.

But anybody looking at species diversity, varieties of successful ecological strategies, and sheer number of individual organisms could conclude that, over the last 400 million years, Earth has become the planet of the insects.

There are more than a million described species in the class of insects, Ectognatha. This class and the less expansive Entognatha (around 5,000 species of arthropods that don't metamorphose) comprise the Hexapods, or everything that has six legs. They outnumber all other species known to science. There are several factors that have contributed to their spectacular success. First, they suffered surprisingly few losses at higher phylogenetic levels during the last mass extinction, the one that defined the end of the Cretaceous and wiped out the non-avian dinosaurs, among many other species. Thus, while much of the biodiversity we see around us, including birds and mammals, has only had around 66 million years to diversify, insects go back more than 410 million years.

In this time frame, there have been several synergistic factors that enabled explosive growth of some groups at later points, without triggering that expansion directly. These include the first development of flight, soon after the arthropod ancestors came out of the oceans, then, at a later point, complete metamorphosis, which led to the Holometabola groups now comprising more than 80% of insect diversity.

Combined with an especially flexible development and a robust and versatile body plan based on the arthropod exoskeleton, these traits enabled insects to shape their world and interact with many other species in numerous different ways.

In the Early Cretaceous (146–100 million years ago), flowering plants (angiosperms) spread around the world, and pollinating insects played an important part in this. Much of the current concern about population declines in insects is because the loss of pollinators may ultimately threaten our food provision.

Over the same timescale, another kind of interaction evolved between many kinds of plants and species of ants, where plants provided food and shelter for the ants, and ants offer defense and seed dispersal for the plants. They even protect plants from diseases.[1] Research has shown that mutualism between ants (family of Formicidae, with more than 12,000 named species) and plants has enhanced diversification of plants, but its impact on the ants has been less clear-cut.

Matthew Nelsen (we already met him in chapter 1) and colleagues at the Field Museum of Natural History in Chicago have addressed this question by analyzing relevant traits in phylogenies covering more than 10,000 plant genera and 1,730 species of ants.[2] Their results suggest a sequential series of events rather than a simultaneous boost for both sides.

Ancestral ant species were carnivores nesting and foraging on the ground. Foraging on plants was the first relevant change, followed by accepting and looking for plant food. Omnivorous ants started to feed on plant sap and possibly on the honeydew produced by aphids. Plant structures beneficial for ants only evolved later. These include extrafloral nectaries secreting nectar on stems or leaves, elaiosomes (which are food packages enticing ants to carry seeds with them), and domatia (hollow structures in which ants can nest, such as the spines formed by some species of acacias). These adaptations enabled ants to become fully engaged in a plant-based niche that largely evolved in the Cenozoic—that is, after the last mass extinction.

Thus, the authors describe a stepwise increase in mutual dependence. During this tightening of the relations, the plants showed enhanced diversification rates, but the ants did not.

To expand these studies to a wider range of species, Katrina Kaur and colleagues at the University of Toronto, Canada, used text mining of more than 89,000 published abstracts to capture published traits associated with ant-plant mutualism.[3]

Their analysis suggests, as Nelsen and colleagues also found, that ants diversified before becoming engaged in mutualistic relations with plants. "To our surprise, the intimate and often beneficial relationships that ants have with plants apparently did not help to generate the over 14,000 ant species on Earth today," Kaur said. "Mutualism may put the brakes on the rise of new species or increase the threat of extinction because an ant's fate becomes linked to its plant partner's."[4]

While ants developed many ways of living with and on plants, the group of leaf and stick insects (order Phasmatodea) evolved sophisticated ways of resembling plants in an effort not to be detected by predators hunting by vision. A relatively recent radiation has led to more than 3,000 species that may be very diverse visually but are hard to sort into a phylogenetic tree based on existing sequence data. Therefore, the deeper connections of their evolutionary history have remained controversial.

Sabrina Simon from Wageningen University and Research in the Netherlands and other researchers from around the world analyzed 27 new transcriptomes of Phasmatodea in combination with previously published sequence information.[5] For each species, more than 2,000 separate genes were analyzed.

The results contradict some of the earlier divisions made based on appearances. Instead the new phylogenetic tree has a strong geographic division into western ("New World") and eastern members of the order that extend from Africa through to New Zealand. Appearances proved deceptive on the island of Madagascar as well, where a flamboyant diversity of stick insects descended from a single ancestral species that colonized the island some 45 million years ago.

The oldest lineages of stick insects lead back to the time after the last mass extinction, 66 million years ago. This finding suggests that the characteristic camouflage evolved in response to the rapidly growing threat of birds and mammals taking over after the dinosaurs.

Insect orders have been shaken up and redefined in the last few decades as molecular studies forced revision of classifications based on anatomy and behavior. As recently as 2007, molecular research confirmed that termites are just a eusocial group of cockroaches. While this relationship had been suspected since the early twentieth century, it was not reflected in traditional systematics. Losing their status as a separate order, the termites were merged into the order Blattodea, which now includes around 4,400 species of cockroaches as well as more than 3,000 species of termites.

Many of these species play important roles as degraders of dead wood, which also makes them noticeable players in the global carbon cycle and thus in climate regulation. Some termite-derived enzymes are of interest for the production of biofuel from otherwise indigestible biomaterials. Some Blattodea are considered pests, and termites, in particular, can endanger wooden structures in some areas. Thus, there are many ecological and economical reasons to better understand their biology.

In 2019, Dominic Evangelista from the Sorbonne in Paris, France, and colleagues presented a new, detailed phylogenetic analysis based on 2,370 protein-coding genes from 66 species representing all but one major groups of Blattodea.[6] Their results confirmed the affiliation of termites within the cockroaches and clarified some long-standing problems, such as the age of the order. While previous work had suggested origins some 240–180 million years ago, long before fossil evidence can be linked to it, the study suggests that most major groups of Blattodea arose during the Cretaceous (147–66 million years ago), which is more easily reconciled with the fossil record. It also proves that any fossils resembling cockroaches in the Carboniferous (359–299 million years ago) can be safely excluded.

Social behaviors like maternal brood care are not ancestral features of the group, the researchers found. Instead, they suggested that the common ancestor packaged its eggs in a protective parcel, the ootheca, and buried this in a hole, as many cockroaches still do today. This behavior may also explain the evolution of wood degrading, the authors said. They conclude the paper by saying, "We suggest that the transition from a free-living ancestor to living within wood (in *Cryptocercus* and termites) could have been partially driven by advantages gained in protection of the ootheca."

Zooming in on termites, and on the family of Termitidae in particular, Ales Bucek and Thomas Bourguignon from the Okinawa Institute of Science and Technology Graduate University, Japan, describe the origins of fungiculture by termites.[7] Termites of the subfamily Macrotermitinae produce comblike structures in their hives where they cultivate fungal symbionts. This, the authors find, is a derived feature probably unrelated to the earlier loss of protozoan symbionts in the termite gut.

Some of the orders of insects have been a challenge to systematics, but the very root of the insects' family tree poses even bigger and more fundamental questions. A key event is the first development of flight in animals, which separated the Pterygota (winged insects) from the likes of silverfishes (order Zygentoma) and bristletails (order Archaeognatha).

Early flying insects had stiff wings set at a fixed angle, like in today's mayflies, damselflies, and dragonflies. The evolution of foldable wings defines the group of Neoptera, which covers the Holometabola (fully metamorphosing insects) as well as the Polyneoptera (e.g., stick insects, roaches, and termites) and the Paraneoptera (e.g., true bugs, lice, and thrips).

The fossil record features a major gap after the oldest known insect from 411 million years ago, so researchers only have genetics and extrapolations from more recent finds to go by.

Benjamin Wipfler from the Zoological Research Museum Alexander Koenig in Bonn, Germany, together with Sabrina Simon at Wageningen and others, reported a large-scale phylogenomic analysis of Polyneoptera based on more than 3,000 protein-coding genes sampled from more than 100 insect species. Their study aimed to elucidate the evolution of 112 traits, including diet, social behavior, and habitat of larvae and adults, as well as the origins of flight.[8]

On the very origins of the first winglets that led to the evolution of flight, there have been several competing hypotheses, including one that involved insects gliding from trees to the ground, along with two hypotheses based on presumed aquatic stages, as can be observed in today's dragonflies. Wipfler and colleagues concluded from their findings that they can rule out an aquatic lifestyle for the common ancestor of winged insects. Thus, out of the three hypotheses, their results favor the glider one. This also fits with their finding that the common ancestor did not have anatomical adaptations typical of ground-living insects, such as a flattened underside, so it was likely to have spent at least some of its life on plants.

The data also suggested that some behaviors and lifestyles seen in several groups of insects, including social behaviors and life on plants, have evolved several times and are not ancestral. The same holds for the anatomical adaptation of ground-living insects. By contrast, the characteristic arrangement of biting mouthparts below the head capsule (orthognathy) is likely ancestral and has been more widely retained in Polyneoptera than in other groups.

Despite this unrivaled success story lasting more than 400 million years, ecologists now have reason to worry about the declines observed in some orders, such as the Lepidoptera (moths and butterflies, see below), and pollinating insects. Public perception and economic interests give prominence to bee problems, even though the honeybee is just a domesticated species that humans can propagate as necessary.

The observed losses in pollinators may, however, be just an indicator of wider extinction threats caused by the use of pesticides and the homogenization of the environment. These rapid changes introduced by humans drive up extinctions and reduce the opportunities for speciation, undermining the long-running recipe for success of insects.

Because insects are so widespread and crucial in virtually every terrestrial ecosystem, their demise would have dramatic repercussions for life on Earth in general, not least of all for the survival of our own species. In this situation, on the verge of a possible human-made mass extinction, it is all the more important to understand how the diversity evolved and how it survived previous crises. In short, as we live on the planet of the insects, we'd better understand how it works.

How insects shape our world

Insects generally have a bad reputation. With the exception of domesticated honeybees and beautiful butterflies (but not their very hungry caterpillars!), humans tend to regard the arthropods as nuisances, pests, destroyers of our food sources, and disease carriers. More than half a century since Rachel Carson (1907–1964) warned of a "silent spring" due to the use of pesticides, the killing has become somewhat more sophisticated, but collateral damage is still substantial.

Neonicotinoids were developed in the 1990s as a new type of systemic pesticide that would only kill those insects that eat from the treated plant. That was a brilliant and promising idea at the time, but when it was put into practice, accumulation in the environment and subtle low-level effects in social insects like bees have led to serious concerns over their impact on pollinators and the consequences for insect eaters like birds.

Serving as food may be the biggest role that insects play in the global biosphere, but their tireless activities also help to cycle nutrients, keep soil fertile and aerated, and to make the most of available water resources, as some recent discoveries have demonstrated.

Termites (infraorder Isoptera), for instance, generally get bad press due to the damage some of their species can do to wood structures. Among the over 3,000 species identified so far, under 200 are considered pests, mainly due to their ability to digest wood and to hollow out timber structures until they collapse. This, in combination with the dramatic population dynamics typical in many insects, can lead to problems in the tropical and subtropical climate zones around the globe. Structural damage caused by termites in the southwestern United States has been estimated to be several billion dollars annually.

However, termites' ability to digest wood can be regarded as a very attractive feature in the context of our attempts to produce sustainable biofuels. In fact, the termite gut has served as inspiration for experimental systems designed to degrade wood and other cellulose fiber waste and turn it into fuel. In many ecosystems around the world, the cellulose-degrading ability of termites is an important part of the food web.

The architecture of termite mound structures in arid environments can also serve as an example to humans, as the clever use of aeration channels and the orientation of the mounds helps to keep the inside temperatures as constant as possible. The Australian compass termites (*Amitermes meridionalis*) are famous for adjusting the long axis of their mounds, which can grow several meters high, exactly in a north-south direction, such that they absorb more warmth from the Sun in the morning and evening, and less of it at noon.

Other benefits of termite mounds have only been discovered more recently. Ecologists working on the Kenya long-term exclosure experiment studied the patterns that typically arise when semi-arid landscapes, like the African savanna, dry out and gradually turn into deserts.[9] Characteristically, the surviving vegetation clings on in regularly spaced patches, which effectively hold on to the last available water resources at the cost of the areas in between, which are left to dry out completely. After the initial analyses focused on the effects of grazing animals like elephants and cattle, satellite images made the researchers realize that termite mounds play an important role. They may even account for some of the regular patterns previously attributed to feedback in the use of water and other resources.

The group of Corina Tarnita at Princeton University modified an established model of such feedback processes by adding in termite mounds such as those built by the fungus-cultivating species *Odontotermes* in Kenya, which look like lens-shaped humps on the ground but have extensive chambers and passages below ground.[10] The modified model could accurately reproduce the existing distribution observed in Kenya. An important insight from the study is that the regular patches hitherto taken as an alarm bell signaling the threat of complete desertification, may on the contrary signal the presence of termite mounds, which can confer resilience to climate change.

Simulating a prolonged drought in their model, Tarnita and colleagues predicted that the termite mounds, by improving the availability of the scarce water, can enhance resistance to drying out and speed up recovery after a dry period. The authors noted that their model applies on the timescale of years to decades but may fail if a climate disruption becomes permanent on a longer timescale such that the plant and insect species involved may no longer be viable in the area.

In a follow-up work, Tarnita, together with Juan Bonachela from the University of Strathclyde, United Kingdom, developed a more advanced model designed to fully cover the co-operation of natural feedback patterns and eco-engineers like the termites.[11] With this improved model, the authors provided a detailed explanation of the "fairy circles" (conspicuous round patches with no vegetation that give the impression that plants form a ring around them) in the Namibian desert. Like other regularly spaced patterns found in nature, these patches have fueled long-running debates between proponents of a feedback model and those of an insect-based model.

The competition between large termite states, which expand and exterminate smaller rivals until they reach a stalemate with an equally mature neighbor, explains the hexagonal geometry of the patterns, in which most patches have six equally distributed neighbors. "Eventually you end up with colonies of very similar sizes that are as far from each other as possible, while at the same time not leaving any space unoccupied," Tarnita said.[12] At the same time, the feedback part of the model can account for the arrangement of plants in the spaces between the patches, including a smaller-scale clustering that previous models did not cover.

There are differences between patterns forming in different environments—for instance, in Kenya the termite mounds are vegetation hotspots, while in Namibia they show up as bare patches, but the geometric principles that lead to regular patterns in the first place are quite general, such that the mathematical models used by Tarnita and colleagues can also be applied in other ecosystems. "One of our goals in this work was to understand how vegetation patterns can form from territorial competition for resources between social insect colonies, but the model could apply very broadly to characterize spatial patterns in other territorial animals," she said.

The termites in Tarnita's Kenyan study grow fungi in their mounds, a feature they share with many ant species, although these are in a different order of insects (Hymenoptera, together with bees and wasps) and the similarities are likely down to convergent evolution.

In either case, the procedures of insect fungiculture require task division and a highly sophisticated set of behaviors, including not only the spatial organization of their environment but also measures of hygiene and waste disposal. In a recent study on leaf-cutting ants, Andrés Arenas and Flavio Roces from the University of Würzburg, Germany, analyzed the information transfer processes by which foragers of an ant colony learn to avoid certain kinds of plants if they prove detrimental to the fungi they cultivate.[13]

The researchers conducted experiments on laboratory colonies of the species *Acromyrmex ambiguus*, which builds special chambers for waste disposal. The researchers treated the ants' favored plant material with fungicides, which affected the cultivated fungi and forced the ants to dispose of the unsuccessful cultivars and change their foraging preference. Then, by switching waste chambers between colonies that had and those that hadn't been exposed to the fungicide, the researchers could show that the materials present at the waste dump conditioned the ants to avoid the type of plant that was affected, even if they didn't experience the cultivation failure.

While some ant species dispose of their waste outside, the more labor-intensive use of purpose-built waste chambers appears to be an advantage in that they serve as an external memory of failed cultivation attempts. In line with this explanation, the researchers could show that the timescale on which the avoidance persists is comparable to the lifetime of the waste chamber.

Ants also share the termites' ability to influence the water storage properties of the soil they live in, potentially improving the soil quality for plants and making the ecosystem more resilient to change. In 2017, Ming'an Shao from the Chinese Academy of Sciences in Beijing and colleagues systematically investigated how the aggregate mulches produced by the burrowing ant species *Camponotus japonicus* affected the water retention and temperature of the soil.[14]

When ants excavate their underground structures, they break up soil particles with their mandibles and move the fragments to the surface, effectively covering the surface with an aggregate mulch. The particle size depends on the size of the species and its mandibles. The species studied by Shao and colleagues has a large body, at up to 12 millimeters (0.5 in.) in length, and produces particles of around 1.6 millimeters (0.6 in.) in diameter. By moving the ant mulch to soil with controlled conditions, the researchers conducted experiments allowing them to analyze how the soil characteristics are affected by it.

The investigation showed that ant-produced mulch layers of up to 9 millimeters (0.35 in.) thickness were efficient at keeping moisture in the ground and blocking sunlight from heating it during the daytime. This study was conducted in the context of large-scale landscape alterations in the Grain for Green project on the Northern Loess Plateau in China, which was designed to control the severe losses of soil to erosion. By covering the ground with their aggregate mulch and thus helping vegetation, the ants made their contribution to this effort.

Apart from pollination and soil improvement, the third big role of insects in ecology is to serve as food for other species. Ecologists are acutely aware of the connection between insecticide use and bird declines, but birds aren't the only animals to feed on insects. In warmer climate zones, widespread termite mounds attract specialized mammalian predators, such as the aardwolf (*Proteles cristata*), and even our closest living relatives, the chimpanzees, have been observed using tools to catch termites.

Globally, however, the most proficient predators of insects are spiders. Martin Nyffeler from the University of Basel, Switzerland, and Klaus Birkhofer from Lund University, Sweden, re-assessed the amount of insects consumed by spiders and arrived at astonishing figures.[15]

The authors used two independent routes for the estimates, one based on the standing biomass of spiders, the other on extrapolation of observed killings. The first method arrives at a total spider biomass of 25 million metric tons (28 million tons), and an annual food requirement of 460–700 million metric tons (507–772 million tons). Most of this prey is made up of insects and collembola (springtails)—a subclass of six-footed arthropods that is no longer grouped within the insects. The second route yielded a slightly wider range around a similar midpoint, at 400–800 million metric tons (440–882 million tons).

This is comparable to the meat and fish consumption of the human population, estimated to be 400 million metric tons (440 million tons) in 2003. This also suggests that when we are ready to give up our own unsustainable predation habits (see chapter 5), there will be enough insects around to feed us.

While their specialization on insects could make spiders a useful helper in pest control, the authors note that fields with annual crops provide very poor habitat for spiders and a short feeding season. Therefore, these fields have only a small share in spider biomass and in the global tally of insect predation by spiders. Nevertheless, in certain crops and with a practice of low pesticide use, spiders can make a significant contribution to pest control.

Such surprising discoveries from the worlds of arthropods remind us that our understanding of the natural world—which we are currently destroying at an unprecedented speed—is still woefully incomplete. Complex networks of interactions between underappreciated organisms may help our efforts to feed a growing world population in sustainable ways, while our simplification of the environment by use of monocultures protected by pesticides may undermine our best efforts. Many insect species have been eco-engineers for millions of years, and in our quest for a more sustainable way of life, we can still learn more about and from them.

Beneficial bugs

Bug is a word we use for a multitude of unrelated things, including technology malfunctions, hidden microphones, viruses, microbes, and all sorts of animals that happen to be small. When we use it as a biological term, bugs must be insects, but even within this class, confusion reigns, as the American English term *ladybug* refers to a beetle, and *lovebug* to a fly.

True bugs are, depending on your creed, either members of the order Hemiptera, or, within that order, the suborder Heteroptera. With over 75,000 species, the Hemiptera include a wide variety of forms, which share one defining anatomical feature, the tubular structure that most, like aphids, use to suck saps from plants and some, like bedbugs, use to feed on other animals, including humans.

These highly specialized mouthparts for sucking liquid nutrition from plants or animals evolved more than 300 million years ago in the shared ancestor of the Hemiptera, which then rose to become the predominant group of insects in the Permian, between 300 and 250 million years ago. Only after the Permian mass extinction (which marks the boundary between Permian and Triassic, 252 million years ago), they were overtaken by those orders of insects that undergo a full cycle of metamorphosis, including the beetles, bees, and butterflies, collectively known as the Holometabola.

We marvel at the fragile beauty of butterflies, appreciate the pollination services of bees and worry about the threats they face, and admire the skills of ants, but bugs generally get less appreciation. Even their position in the tree of life is still subject to debate.

The systematics of the hemipteroids, the true bugs and the insect groups related to them, has seen a few upheavals. This larger grouping also includes bark lice and parasitic lice (order Psocodea) and thrips (order Thysanoptera). In 2014, a phylogenetic analysis across the insect class suggested the Psocodea should be divorced from the bugs and grouped as a sister taxon with the Holometabola, a suggestion conflicting with most previous classifications.[16]

Kevin Johnson from the University of Illinois Urbana–Champaign and colleagues addressed this issue with a major analysis of nearly 2,400 protein genes, their transcripts, and their predicted amino acid sequences across 193 insect species. Although one type of statistical analysis of the data obtained appeared to be compatible with the 2014 re-assignment, other analyses came out in favor of the old order, keeping both lice and thrips together with the true bugs.

In another controversial question, the research confirmed the 2014 analysis, which had fewer data points. Both agreed that, within the order Hemiptera, Coleorrhyncha (moss bugs) and Auchenorrhyncha (a group including cicadas, leafhoppers, and their relatives) are sister clades.

The results also allowed the researchers to date the evolution of the specialized tubular mouthparts that define Hemiptera to at least 350 million years ago. Where other insects relied on biting off little bits of material to feed, this evolutionary step enabled the bugs to find ecological niches sucking liquids out of plants or animals.

Further insights into the vast diversity of bugs and other insects are bound to emerge from the Earth BioGenome Project, which aims to obtain the genome sequence of every eukaryotic species known to science. As mentioned in the introduction to this chapter, the project will have to spend significant parts of its resources on the vast species richness of insects.

Many bugs are seen as pests, which probably explains the use of the word for anything small and undesirable. Part of the problem is their ability to reproduce very rapidly if food is provided. Many aphids, for instance, can use asexual (parthenogenetic) reproduction to rapidly boost numbers.

Cicadas have become proverbial as insect plagues, as some species have evolved the strategy of emerging as vast swarms in long-term cycles (e.g., with 13- or 17-year intervals).

To the ecologist on the other hand, many of these plagues are just the bottom layer of the food web, feeding on plants and in turn providing food for other insects and birds. Aphids, as every gardener appreciates, are consumed and controlled by ladybugs.

Some kinds of aphids are also farmed by ants, in a symbiotic relationship that is built on the carbohydrate-rich honeydew that aphids exude. This nutritious excretion is the consequence of the fact that the saps they suck from plants contain excessive amounts of carbo-hydrates in relation to the amount of protein the aphids need.

A few species of hemipterans are valued as pollinators. In North America, for instance, the milkweed bug (*Lygaeus kalmii*) is known to feed on the nectar of various herbaceous plants and thought to be helping their pollination. In a study of pollination in *Maca-ranga* (Euphorbiaceae) on Amami and Okinawa Islands, Chikako Ishida from Kyoto University, Japan, and colleagues could show that bugs are the main pollinators of these plants.[17]

The researchers found that the flower bugs *Orius atratus* were the predominant visitors to both stamina and pistils of *Maca-ranga tanarius* on both islands. Detecting body pollen on *O. atra-tus* collected from pistils, the researchers concluded that the bugs moved from staminate to pistillate trees, thus contributing to pol-lination, and that they were the major pollinator of this tree species.

Intriguingly, the tree protects its nectar in a ball-shaped struc-ture that bugs can pierce with their sucking organ, while other kinds of insects cannot. Apart from the nectar, the bugs also feed on thrips living on the plants.

Other remarkable features and abilities that evolved among the bugs include the noisy songs of the cicadas and the ability of vari-ous freshwater species to walk on water, using the surface tension for support, or even to drive on the surface using the same mecha-nism as a detergent-driven toy boat.

Some of the predatory bugs are somewhat harder to defend, especially those that have specialized on humans, like the bed bugs (genus *Cimex*), and may transmit diseases, like the kissing bugs (subfamily Triatominae), which tend to feed near people's mouths when they sleep and which transmit the protozoan pathogen of Chagas disease, *Trypanosoma cruzi*. Still, there are various ways in which bugs can be helpful to humans.

Historically, a few species of bugs have found singular uses. Thus, the carmine dye produced by the cochineal (*Dactylopius coccus*), a scale insect from the suborder Sternorrhyncha, was valued in the Mayan and Aztec empires as well as under colonial rule.

D. coccus lives on prickly pear cactus (genus *Opuntia*) in the tropical and subtropical regions of the Americas and produces carminic acid as a deterrent against insect predators. This complex organic molecule accounts for up to a quarter of the insect's dry weight, such that it can be readily extracted and used as a textile and food dye. In European antiquity, it was a different species of Sternorrhyncha that served the same purpose, namely *Kermes vermilio*, which was used in the production of crimson.

Some hemipterans are already used in pest control. These include obligate predators such as the Nabidae (damsel bugs), as well as omnivores such as *Dicyphus hesperus*. Another species used in pest control is the spined soldier bug (*Podisus maculiventris*), which feeds on the larvae of the Colorado beetle and the Mexican bean beetle.

Some Hemiptera are traditionally eaten in parts of Asia and Africa, including cicada in China as well as in Malawi. Considering the continuing global growth of meat production that is unsustainable in terms of both land use and climate change, a much wider appreciation of insect food would be desirable. Research is needed to develop insect-based nutrition that is attractive to consumers currently used to a diet rich in beef.

Bugs tend to produce a range of remarkable compounds that help them to retrieve nutrients from plant or animal targets in a liquid form while evading the prey defenses and deterring their own predators. These include antimicrobial peptides and immune-modulating compounds, which should in principle be of interest for medical applications, but early attempts at utilizing bug chemistry commercially have failed to bear fruit.

One promising target for bug-derived antibacterial substances is malaria. Andreas Vilcinskas's group at the University of Giessen, Germany, demonstrated for the first time the anti-plasmodial activity in vivo of a defensin from the tick *Ixodes ricinus* tested in a mouse model of malaria.[18]

Defensins are antimicrobial peptides, meaning that they are not suitable for oral medication, as peptides would immediately be digested. However, the results suggest that injection of this tick defensin could become a useful way of treating malaria cases, especially those with plasmodia resistant to other treatments.

Vilcinskas, who also chairs a research center devoted to "yellow" (i.e., insect-derived) biotechnology, investigates a range of avenues for other insect-derived bioactives, including medical maggots for wound treatment and mealworms for protein-rich food and animal feed.

As the ongoing human-made destruction of the biosphere accelerates, it is becoming more urgent to recognize all its parts and try to avoid destroying species that may hold the key to our own future success. Technologically speaking, bugs aren't bugs in the global system. They are a feature, and the system might not work without bugs.

How locusts become a plague

Locust swarms devastating entire landscapes have affected human societies since the beginnings of agriculture. Their sudden appearance and destructive effects are described in ancient texts, including the Bible, the *Iliad*, the Qur'an, and the Mahābhārata.

In 2020, as the world was kept busy by an outbreak of a different kind, locusts swarmed in East Africa and the Middle East, affecting countries including Kenya, Ethiopia, Uganda, Somalia, Eritrea, Iran, Yemen, Oman, and Saudi Arabia, and reaching even

Pakistan and India. Meanwhile, an upsurge of South American locusts that began in 2015 and surpassed all outbreaks seen in the previous 60 years continued in Argentina, Bolivia, and Paraguay.

Some parts of Kenya saw the worst devastation by locusts in 70 years. As the swarming behavior is the desert locust's evolutionary adaptation to the sudden change in conditions when abundant rainfall follows a drought, extreme weather events facilitated by climate change may be a contributing factor in the catastrophic event.

Systematic monitoring of the conditions that may lead to outbreaks can enable early control measures that stop the situation from escalating. In politically unstable parts of the world, however, these measures may be hindered by security concerns. When monitoring fails, as it did in late 2019, and devastating locust plagues threaten to cause food shortages for many millions of people in East Africa and beyond, large-scale spraying of organophosphate insecticides is used to decimate the swarms. As a report by the environmental website Mongabay highlighted in April 2021, the products purchased by the Food and Agriculture Organization (FAO) to combat the plague are mostly insecticides considered harmful, including chlorpyrifos and fenitrothion, which are listed as "moderately hazardous" by the World Health Organization.[19] Their application risks causing additional damage to the environment and the resident population.

A biological control agent based on spores of the fungus *Metarhizium acridum*, which infects and kills locusts, is also available. It has the immediate advantage that it affects a much narrower range of species, although there are concerns for some non-target species, such as termites. Its use can be hindered by a lack of country approvals, production volumes, and economic feasibility. As a biological material, it has a shorter shelf life than the chemical agents, and its production cannot be scaled up as rapidly. Related to the 2020–21 outbreak in Africa, the FAO and the governments of the affected countries purchased more than 12,000 liters (3,200 gal.) of fungal products—which corresponds only to a hundredth of the volume of organophosphates acquired.

As locusts only become a problem when they switch from the solitary to the gregarious, swarming lifestyle, it is vital to better understand that switch and find ways of preventing it. Over the last 100 years, much research has been dedicated to the phenomenon, but many questions remain.

Until the early twentieth century, swarming locusts were considered a separate species from the solitary specimens. Only with the work of Boris Uvarov (1889–1970), a Russian émigré who worked in London, United Kingdom, it became clear that certain grasshopper species, such as the desert locust, undergo a dramatic change in phenotype when conditions are favorable. While locusts are not a systematic group, only grasshopper species that undergo a swarming phase are described as locusts. This coexistence of different phenotypes, or phases, is known as density-dependent polyphenism. The phase change from the solitarious to the gregarious phase may be triggered by a serotonin signal and maintained by the exchange of pheromones, which depends on population density. It affects the appearance and the behavior to an extent that they really do look like different species.

Under suitable conditions, gregarious locusts reproduce abundantly, giving rise to bands of wingless nymphs, which later form larger swarms of adult insects. They can pass on the gregarious phenotype to their offspring. Given that rapid reproduction is a big part of the problem, Amir Ayali at Tel Aviv University, Israel, argues, the mating behavior of these insects is surprisingly little studied. Only recently, Ayali's group reported experimental work with desert locusts (*Schistocerca gregaria*) in the gregarious as well as in the solitary phase.[20] Ayali summarized this conclusions as follows: "We suggest that intra- as well as inter-phase reproductive interactions among gregarious and solitarious locust populations have a major impact on the locust population dynamics."[21] Out of the four possible combinations, the researchers found that the most successful ones were gregarious males mating with solitary females. The offspring typically inherits the phase status of the mother, although the mechanisms of the transmission are yet to be fully understood.

Another underappreciated factor in the environmental plasticity of locusts is the role that their bacterial symbionts may play. In a review of the field of locust microbiota, Ayali and colleagues raised the possibility, still unproven, that the gut bacterium *Weissella cibaria*, found to become more dominant in gregarious locusts, may contribute to the phenotype switch.[22]

They may even be involved in passing on the trait, as the microbiota is transmitted from mother to offspring via the egg-pod's foam plug, through which the larvae pass upon hatching, as Ayali's group has shown.[23]

In the first systematic study of the microbiota of the reproductive tract in gregarious and solitary locusts, Omer Lavy and colleagues from Ayali's laboratory found that the microbiota is changeable with population density and thus could play a role in the polyphenism.[24] Earlier work had shown that they may also be involved in production of the pheromones that control the phase status.

In addition to the behavior changes induced by the phenotype switch, Ayali's group found that the experience of collective walking triggers a change of motion status that can be observed even when the individual insect walks on its own.[25] For this study, the researchers used desert locusts grown in the lab consistently held in the gregarious phase because of the density of living and availability of food. These insects would not normally have experienced collective movement, but if they are exposed to it, the researchers observe the switch to different walking kinematics, which individual insects upheld when they were removed from the crowd.

The capacity for demonstrating collective motion, including the "collective-motion-state," is part of the gregarious locust phenotype and thus does not need to be learned. Ayali suggested that the state is released upon experiencing collective motion. His group tried to induce the same change in solitary animals without success, suggesting that the capacity depends on elements of the gregarious phenotype.

This propensity for collective behavior is useful for swarms, as it means an individual accidentally separated from the swarm will find it easier to fit back in when it catches up. Thus, while each of the phases defines a different set of behavioral patterns, there are further complexities within each set that must be taken into account.

The definition of a separate movement state better suited to engaging in collective behavior than the naïve state challenges the established view of collective behavior as an emergent property of many individuals without calling for a specific status change. The physiological foundations of this change and its long-term fate remain to be explored.

The research also showed that the collective behavior depends on very little interaction between individuals, which makes it even more impressive that the bands and swarms are tightly co-ordinated on a large scale.

Gregarious locusts can switch back to the solitary phase when they have been isolated for an extended period—the timescale of this reversal varies widely between species. While this process can be induced in the laboratory, it doesn't offer a solution for the problems posed by swarms of billions of insects. Once in the open, that genie won't go back into the bottle if there is any food left for it to eat.

Beyond the collective behavior, the energy balance of swarming locusts may be an important handle to limit the damage they cause. Arianne Cease and colleagues at Arizona State University (ASU), Tempe campus, and at the Chinese Academy of Sciences showed that locust swarms prefer vegetation with a low protein content, suggesting they focus on stocking up on carbohydrates as fuel for their energy-intensive migrations and swarm behavior.[26] In practical terms, this means that heavily grazed fields, which have a high proportion of carbohydrates to protein, promote locust outbreaks, while crops grown in healthy soils with a high nitrogen content are less affected. The researchers found this connection confirmed around the globe, with different locust species on different continents.[27]

In an experimental study building on that discovery, Marion Le Gall and colleagues from ASU and from Senegal exposed locusts to millet plants (*Pennisetum glaucum*) that had received high, moderate, or no nitrogen fertilizer.[28] The researchers could show that the fertilizer treatment increased the protein-to-carbohydrate ratio in the plants. When Senegalese grasshoppers (*Oedaleus senegalensis*), which evolved the locust traits independently from the desert locust and to a less extreme extent, were locked in with the highly fertilized plants, they showed reduced survival and reproduction rates compared to those on unfertilized plants. Another experiment led by Stav Talal and colleagues from ASU and from South America showed that marching South American locusts were limited by plant carbohydrate content.[29] High-carb, low-protein diets supported higher survival, growth, and lipid accumulation.

Arianne Cease and Rick Overson at ASU jointly lead the Global Locust Initiative (GLI). Launched in 2018, this initiative consists of the GLI Lab and the Global Locust Network. The network's purpose is to forge new partnerships that seek to better integrate locust research and management. It supports more than 400 members from

55 countries. The GLI Lab hosts projects spanning natural and so-cial sciences that aim to advance the fundamental understanding of the systems in which locusts exist; it also works to help develop local solutions to the global challenge of locust plagues.

In a joint project with Senegalese farmers, USAID (the US Agency for International Development), Gaston Berger University, McGill University, CIRAD (Centre de coopération internationale en recherche agronomique pour le développement), and the Sene-galese Plant Protection Directorate, the GLI Lab is developing a novel long-term and community-based preventive strategy in West Africa. This involves setting up village-based soil amendment and locust monitoring programs to create environments that discour-age locusts and outbreaks of grasshoppers.

The GLI is part of the National Science Foundation–supported Behavioral Plasticity Research Institute (BPRI), which aims to connect all aspects from molecular studies through to ecology on the landscape scale. BPRI is led by Baylor College of Medicine and Texas A&M University and includes partners from four additional institutes. As Cease and her colleagues have highlighted, locusts are part of a complex social-ecological-technological system. The social context, such as governance, often receives less attention but is also critical.

In various countries around the world, such as Saudi Arabia, people enjoy eating locusts. In terms of the efficiency of turning primary plant production into calories for human consumption, these insects outperform domestic cattle more than fivefold. When a locust plague strikes, however, human consumption doesn't quite scale with the sheer biomass bulk of the phenomenon. Moreover, human consumption wouldn't be advisable once insecticides have been deployed.

Keeping an eye on the conditions that might encourage swarm-ing is currently the only way to stop plagues from happening. In

Africa, in particular, extreme changes between drought and excessive rainfall, made more likely by climate change, may conspire with political instability (which hinders monitoring of the early stages) to make catastrophic locust plagues more frequent. The fact that these outbreaks are connected to war and extreme weather aligns with the biblical reports of different kinds of plagues arriving in sequence like a divine punishment.

As Ayali pointed out, nobody wants locusts to become extinct, as the North American species, the Rocky Mountains locust (*Melanoplus spretus*), did quite surprisingly in the early twentieth century, most likely due to land-use change destroying its breeding grounds. "We would much rather keep them solitary," Ayali concluded. "If only we knew how."[30]

CHAPTER 4

Looking After Our Forests

After meeting some of the players in the globally connected network that is the biosphere of our planet, we now come to the way in which we humans interact with the network, and we shall see that it is unsustainable in many ways. After many centuries of treating natural resources as unlimited and indestructible, humanity urgently needs to shift its modus operandi from ruthlessly ransacking the Earth to living within its ecological web and using its resources sustainably.

Forests provide a prime example of what we have done wrong in the past and need to do differently from now on, lest we lose the planetary lungs that allow us to breathe. Ancient civilizations led the way by destroying their natural forest environments to obtain land for agriculture, as well as wood for construction of homes and ships. Early modern societies followed in their footsteps, cutting down ancient forests across Europe, for instance, and draining wetlands. Nowadays, we think of the Amazon as the main forest ecosystem that we have left to lose, but in terms of global climate justice, restoring lost forests (e.g., in Europe) should be as much of an issue as saving the ones that are left in the developing world.

Moreover, a forest is more than a collection of trees. Natural forests are important havens for biodiversity, so replacing them with a monoculture of commercially useful trees won't do much good. We need ecological understanding of what we are about to lose in the Amazon so we can save it there and restore forest ecosystems elsewhere.

The rise and fall of global forests

Ginkgo trees (*Ginkgo biloba*) are now a familiar sight in botanic gardens around the world and in the streets and private gardens in many cities. In the Western world, they are seen as exotic and associated with origins in China, but their back story is much more interesting and complex.

The fossil record shows that in the Jurassic (200–145 million years ago) there were several ginkgo and similar species, producing large, solitary seeds in a characteristic way that is distinct both from conifers and from the later innovation of flowering plants (angiosperms), whose descendants now dominate our parks and gardens. Ginkgos were still thriving in temperate climates around the globe during the Cretaceous, offering shade to the dinosaurs in a world that was much warmer than ours.

They survived the mass extinction at the end of the Cretaceous, but the cooler climate heralded by the appearance of the Antarctic ice shield some 35 million years ago challenged their way of life. As the glaciations brought by recurring ice ages kept pushing back vegetation, ginkgos had to reclaim the higher latitudes each time. After the last ice age, they failed to do so and disappeared from most of their former habitats, except for a part of China. Some experts believe that an animal species helping to disperse ginkgo seeds had become extinct, such that the species could only hold the area it occupied already but could no longer move with shifting climate zones.

This might have been the death knell for ginkgo, but ancient Chinese culture cherished the tree, and humans spread it within China and then around the world again. Amid the thousands of species that *Homo sapiens* has driven into extinction, here is one that we probably saved from that fate, long before we had any concept of extinction or conservation.

It goes to show that trees, although immobile as individuals, may have to move to different territories as populations in order to survive in the long term. As human-made climate change is shifting climate zones more rapidly than most natural events have in the past, the mobility of tree species becomes an important part of their resilience to change, and it is therefore vital to understand how and why tree species came to the places where they are thriving now, and how they can adapt when their survival is threatened.

Tropical forests contain the most abundant plant diversity. As they continue to decline due to deforestation for agricultural use and fragmentation by infrastructure measures such as roads, researchers are now rushing to understand this diversity and its evolutionary origins while they can.

In a major new assessment of phylogenetic relations between the species in the global tropical forests, Ferry Slik from the University of Brunei Darussalam and Janet Franklin from the University of California, Riverside, with more than 100 colleagues from around the world, have assembled a standardized dataset of angiosperm trees in old-growth tropical forests at 406 separate locations representing all major regions of tropical forests and the whole range of environmental conditions.[1]

Analyses linking the results to a phylogenetic tree resolved to genus level showed that all tropical forests have a similar mix of ancient tree lineages going back to the Late Cretaceous (100–66 million years ago). At that time, the supercontinent Gondwana had already broken up into today's continents of Africa and South America, but plant dispersal appears to have continued across the nascent Atlantic Ocean.

Only with the kind of analyses that emphasizes recent divergence could the authors establish different patterns for today's biogeographic regions. Based on this analysis, they propose a floristic division of the tropics that differs from traditional divisions, distinguishing the combined American and African forests from the Indo-Pacific ones. Dry forests in America, Africa, Madagascar, and India are grouped together, as are subtropical forests around the globe. The researchers hope that understanding the large-scale structures may help with forest conservation.

In a more close-up perspective, Michael Pirie and colleagues from the University of Mainz, Germany, looked at the diversity and distribution of the genera *Cremastosperma* and *Mosannona* from the Annonaceae, or custard apple, family in South America, which typically form shrubs or small understory trees.[2] Based on genetic data, they determined a phylogenetic tree of 35 species from these genera and related this to the current distribution and the history of geological events—namely, the formation of the Andes, the drying out of the Pebas wetlands system, and the development of a land bridge to Central America.

While the movements of the genera over geological timescales inferred from these analyses brought no major surprises, they did show that the two genera, now occupying overlapping habitats, arrived in their current territories via different routes. During these studies, Pirie and colleagues also discovered five new species.

Spreading across the (moving) continents and climate zones in geological time, plants diversified and evolved different forms and functional traits to adapt to new environments. Big data approaches now enable ecologists to look for patterns in the ways in which thousands of species evolved over millions of years.

Sandra Díaz from Universidad Nacional de Córdoba, Argentina, and colleagues analyzed six functionally important parameters of plants, including height, leaf surface, leaf mass per surface area, nitrogen content, stem mass per volume, and mass of the seeds, for 45,000 species.[3]

They found that these parameters don't combine randomly but cluster along two main axes. Thus, the seed size tends to correlate with the overall height of the plant. In terms of leaf function, there is a continuum ranging from small, light leaves with high nitrogen content to large, heavier leaves with low nitrogen content, representing different metabolic strategies.

Zeqing Ma from the Chinese Academy of Sciences and colleagues conducted a similar analysis for parameters of the plant roots, arguing that these are at least as important for plant survival as the aboveground parts but have so far been neglected in such studies.[4] Similarly, overlooking the invisible parts of plants has also slowed down our understanding of plant communication (see chapter 1).

Ma and colleagues analyzed thickness and nitrogen content of the primary roots (i.e., those most distant from the stem and directly in contact with the environment) in 369 plant species from seven biomes. They found that relatively thick roots were the ancestral state of affairs when plants started to spread from the tropics, where the growth conditions were constant throughout the year and quite often involved a nutrient-rich, warm, and humid environment.

Expanding to higher latitudes with pronounced seasonal changes and less predictable nutrient supplies, the plants evolved thinner roots that can—with a given investment of carbon—reach out further and move more quickly in the search for nutrients. One important parameter that varied with the root thickness is the extent of symbiosis with mycorrhizal fungi (see chapter 1). As the roots become thinner and more agile, they are less dependent on fungal symbionts.

However, the correlation with nitrogen content observed for leaves in the earlier study does not appear to extend to roots. The researchers conclude that "at the timescale of plant evolution innovations of belowground traits have been important for preparing plants to colonize new habitats, and for the rich generation of biodiversity within and across biomes."

Having successfully spread across most of the land surface of the Earth during the last 200 million years, and having diversified according to the ecological requirements and opportunities, forests now face two unprecedented challenges, which are interlinked and both caused by humans: deforestation and climate change.

Deforestation is often described in terms of surface area lost to land-use change. The global figure of forest loss in the last 300 years has been estimated at 35%, and significant losses continue to be a concern in countries like Brazil, Indonesia, and Russia.

Surviving forests may still be severely affected in their ability to fulfill their ecological role and provide ecosystem services if they are fragmented by roads or disturbed by economic activity. Experimental studies have shown that border effects can severely degrade a forest's function, suggesting that continuing fragmentation can be as damaging as actual surface loss.

In a global, in-depth analysis of forest fragmentation, Franziska Taubert from the Helmholtz Centre for Environmental Research in Leipzig, Germany, and colleagues used satellite data to identify 130 million forest fragments on three continents.[5] They found that the power laws governing the size distribution and geometrical properties of the forest fragments, including fractal dimensions, were surprisingly similar on different continents, despite very different land-use context.

The authors used percolation theory, a tool from statistical physics used in materials science, to analyze the mathematical distribution of the fragments. They found that the systems are close to a critical point, beyond which a further loss of forest is likely to cause a dramatic increase in the number of fragments, along with a further reduction of forest sizes. At the upper end of the authors' predictions, the number of fragments could increase by a factor of 33 within the next 50 years, which would dramatically increase the extent of degradation and function loss near the edges.

Complementing the vision of global forests shattered into a rapidly increasing number of small fragments is a perspective article by James Watson from the University of Queensland in Brisbane, Australia, and colleagues appraising the unique and special benefits provided only by intact forests.[6]

The authors list a wide range of benefits that are provided more efficiently by intact forests than by fragmented ones in areas ranging from climate change to human health. Ecosystem services of intact forests that serve both wildlife and humans include carbon storage and sequestration, weather regulation and hydrological services, buffering against natural hazards including fires and floods, and conserving diversity of forest-dependent life from the ecosystem level through to the intraspecies genetic diversity.

In addition, there are benefits specific to humans, including the shelter of surviving Indigenous populations (who, in turn, are often efficient stewards and defenders of their natural environment) and the avoidance of the transfer of zoonotic diseases. In the case of Ebola, in particular, which has its natural reservoir in bats (see chapter 11), recent research has shown that outbreaks in the human population can be linked to deforestation disturbing the natural virus carriers. The COVID-19 pandemic has given us a warning of how big a problem zoonotic outbreaks can become.

Deforestation in terms of loss of forest-covered surface area is addressed in the Paris Agreement and many other international treaties, as well as in the REDD+ (Reducing Emissions from Deforestation and Forest Degradation in Developing Countries) program, which offers financial rewards for developing countries that minimize their forest loss. Watson and colleagues argue, however, that the quality of the surviving forests is at least as important as their quantity, and that intact forests, far from disturbance and border effects, should also be highlighted in treaties and policy decisions.

The quality of the forested area includes the diversity of tree species. Some reforestation projects, including China's large-scale "Green Wall" against the expansion of the northern deserts, have been criticized for creating monocultures that will be sensitive to disease and environmental change, and will be inefficient at providing ecosystem services.

After the dramatic losses of the last three centuries, roughly a quarter of the Earth's land surface is still covered by forests, but Watson and colleagues cite estimates suggesting that 82% of these are degraded in some way. The intact, diverse forests still need better protection than we grant them today. Otherwise some of their majestic tree species could end up as lonely survivors bereft of their ecological context and planted by humans for decorative purposes, just like the ginkgo.

The Amazonian world before Columbus

In the discussion around the ongoing destruction of the Amazonian rainforest, it would be helpful to know what its natural state was before Europeans arrived in the sixteenth century, or even before humans first arrived just over 11,000 years ago. Some discoveries have fed the narrative that ancient civilizations shaped their environment more thoroughly than we imagine, but more recent work suggests that this human influence may have been limited to specific locations.

Humans conquered the Americas from the top to the bottom, crossing the Bering Strait during a glaciation period some 25,000 years ago and then expanding southward. Archaeological finds and ancient genomes suggest that humans arrived in the Amazon more than 11,000 years ago.

The tropical climate of the Amazon doesn't lend itself to the preservation of the evidence, so we may very well be missing out on important information. The limit of the earliest human settlements in South America has been pushed back to around 10,600 years ago by several recent studies conducted in southwestern Amazonia.

José Capriles from Pennsylvania State University and colleagues studied three forest islands—Isla del Tesoro, La Chacra, and San Pablo—within the seasonally flooded savanna of the Llanos de Moxos in northern Bolivia.[7] This vast plain drained by the Mamoré River has a pronounced tropical climate with wet and dry seasons. The area has long been inhabited and managed, but as it lacks natural stone, there are no ruins of buildings available as evidence of past civilizations.

Studying waste deposits and human remains that had apparently been buried intentionally in ways indicative of an incipient settled community, Capriles and colleagues conclude that early Amazonian populations used these forest islands as a base, at least for a large part of the year, until around 4,000 years ago. Abundant traces of charcoal and burned earth show that they also used fire.

"We have long been aware that complex societies emerged in Llanos de Moxos . . . around 2,500 years ago, but our new evidence suggests that humans first settled in the region up to 10,000 years ago during the early Holocene period," Capriles said. "These groups of people were hunter gatherers; however, our data show that they were beginning to deplete their local resources and establish territorial behaviors, perhaps driving them to begin domesticating plants such as sweet potatoes, cassava, peanuts and chili peppers as a way to acquire food."[8]

Whether there is a connection to the populations that started to develop agricultural practices at the same location in the subsequent millennia remains to be seen.

The Llano de Moxos plain features many mounds and forest islands like the ones that Capriles and colleagues studied. Their origins are controversial—many of these landscape features may have been produced by locals trying to keep their belongings dry during the rainy season and the floods. Evidence of human landscape engineering is beginning to emerge.

In a more recent effort, Neil Duncan from the University of Central Florida in Orlando and colleagues studied two 1.5-meter (5 ft.) drill cores from locations around 20 kilometers (12 mi.) apart in the Llanos de Moxos for traces of the changing environment and human activities.[9]

Specifically, the researchers looked for pollen and phytoliths (microscopic silica particles from plant tissue) to indicate which crops were once grown in the raised fields in the area, as well as for diatoms as indicators of wet conditions and charcoal as evidence of humans using fire. Both drill cores start from dry conditions in the deepest layers, and interestingly they shift to wet conditions at different times: one around 3,500 years ago, the other only 2,100 years ago. As a change in the regional climate should have affected both sites equally, the researchers conclude from this finding that the change to wetter conditions points to human use of water resources and engineering of the environment. This reading is also supported by the increase of charcoal residue at about the same time, pointing to the use of fires for cooking. By contrast, if the cores only recorded natural phenomena, natural fires would have become rarer in a wetter environment, and there should have been less charcoal in association with an increase in diatoms.

Phytoliths and pollen indicative of crop plants also become more abundant at the same time—such that all three indicators studied provide coherent evidence of human land management. The conclusion, therefore, is that as early as 3,500 years ago, people manipulated water and fire and changed the landscape to improve their fisheries with weirs and conduct agriculture on raised fields. The earliest evidence of landscape manipulation the authors found even dates to 4100 BCE.

The evidence suggests the early agricultural population in that area domesticated and cultivated crops, including manioc and sweet potato. In a separate study, Umberto Lombardo from the University of Bern, Switzerland, and colleagues described the Llanos de Moxos as an early center of plant domestication, documenting the cultivation of squash (*Cucurbita* sp.) around 10,250 years before present, manioc (*Manihot* sp.) at about 10,350 years ago, and corn (*Zea mays*) at about 6,850 years ago.[10] Intriguingly, the area is also a global hotspot of linguistic diversity, suggesting that the early farming communities settled there were organized in a patchwork of separate small populations without central control.

This area now has a confirmed record of pre-Columbian agriculture, but was this the exception or the rule in ancient South America? One school of thinking suggests that much of the Amazonian land was managed by its inhabitants, like the Llanos de Moxos, until the Europeans arrived. The dramatic population decline after the European conquest let agricultural surfaces fall into disuse and allowed the rainforest to regrow.

An extreme version of this view made headlines in 2019, when researchers from University College London, United Kingdom, concluded that the sudden regrowth of the Amazon rainforest after the population collapse soaked up enough carbon dioxide to contribute to the causation of the "Little Ice Age" observed in Europe in the seventeenth century.

Alexander Koch and colleagues had combined several methods to estimate the pre-Columbian population numbers.[11] They calculated that European contact, with the subsequent conquests and spread of infectious diseases in a naïve population, led to 56 million premature deaths by 1600 and a population decline from 60 million to 5 or 6 million across the Americas. In areas the Indigenous population had previously used for agriculture, this resulted in forest regrowth on a surface of 558,000 square kilometers (215,500 mi.^{2})—an area comparable to the size of France. This, the authors claim, caused the reduction of the atmospheric CO_2 concentration by 7–10 parts per million, accounting for part of the decrease recorded for that time in gas bubbles trapped in ice cores.

While this is a significant and measurable change to global climate, it corresponds to only three years' worth of the current increase in carbon dioxide caused by fossil fuels. The authors argue that this accidental reforestation counts as an anthropogenic change affecting the Earth system even before the beginning of the Industrial Revolution and that it should be considered in the debate over the start date of the proposed Anthropocene. However, it is a very small change compared to what we are causing right now. Also, one could argue that it only reversed the anthropogenic change caused by the introduction of agriculture in South America.

The finding of a large-scale accidental rewilding after 1492 is noteworthy in the context of current debates around colonialism in that it shows off the misrepresentation in the established narrative suggesting European settlers brought agriculture and more efficient resource use to the Americas. And yet, a rewilded area the size of France (without its overseas territories) still only accounts for less than one-fifteenth the size of Brazil or one-eleventh of the Amazon basin. While some of the Amazon rainforest may have re-grown in the last few centuries, other parts are bound to be much older.

Much effort has gone into showing traces of human influence underneath the current vegetation. More recently, researchers have embarked on showing the opposite—Amazonian forests that remained undisturbed throughout the time span of human presence in South America.

Dolores Piperno from the Smithsonian National Museum of Natural History in Washington, DC, and colleagues have managed to find an undisturbed area of native forest by looking at the interfluvial areas far away from the nearest river and not affected by flooding.[12] They analyzed phytoliths and charcoal residues in 10 earth cores taken in soils underneath mature *tierra firme* (non-flooded, nonriverine) forests in the remote Medio Putumayo-Algodón watersheds in northeastern Peru, covering the vegetation history of the last 5,000 years, and compared the results to a tree inventory of the current forest.

Although the charcoal findings pointed to limited use of fire between 2,800 and 1,400 years ago, the species composition of the flora did not change markedly in ways that could be attributed to human interference. Even those palm species that are today hyperdominant in parts of the forests and could theoretically have been introduced by humans turned out to be a constant presence since long before the first evidence of human activity.

In accord with this finding, Frederick Draper from Florida International University in Miami and colleagues found in their comprehensive analysis of the Amazon that the hyperdominance of a few species accounting for more than 50% of individuals is common across size ranges and the entire area.[13] Thus, hyperdominance of a species on its own does not count as evidence for human influence.

Piperno and colleagues concluded from their findings that ancient residents did not impact the 95% of the total surface of the Amazon basin located away from the water courses as much as they did many riverside locations studied by archaeologists. They summarized their findings stating that "our vegetational and fire history from this previously unstudied, remote region joins the increasing body of evidence that deforestation and fires during the prehistoric period and subsequent vegetation recovery upon European Contact were not so widespread and intense as to have contributed significantly to decreasing atmospheric CO_2 levels and the onset of the Little Ice Age."[14]

Moreover, they suggested that the current composition of these remote forests is not mainly the result of human influences. Although they acknowledged the limitation that more subtle manipulations or planting of certain species of fruit and nut trees may have gone unnoticed in the phytolith record, the authors came to the overall conclusion that "the activities of present and past societies in the MP-A [study area] have not strongly altered the community composition and structure of the species-diverse forests over perhaps thousands of years of utilization." They refer to these societies who used the resources of the forest sustainably over many centuries as "a positive force in maintaining forest integrity and biodiversity."

The finding that a sustainable life in and with the Amazonian rainforest is possible aligns with an idea promoted by non-government organizations working to protect the Amazon, namely that Indigenous groups maintaining traditional lifestyles, including more than 100 uncontacted tribes, are the best stewards of the Amazonian rainforest, preserving it both as a functioning ecosystem and as their own home. As organizations like Survival International have highlighted repeatedly, the forests and their people face the same threats, and efficient measures that protect one will also help the other.

Many know the endangered paradise of Amazonia from the work of Brazilian photographer Sebastião Salgado, who in 2021 published a new volume of pictures and opened an exhibition at the Philharmonie de Paris, France, and who runs a family reforestation project at his family farm in Aimorés, Brazil. In an interview with *The Guardian*, Salgado expressed optimism that quintessential wilderness can survive, noting that we have only destroyed "a little bit of the periphery." He went on to say, "The heart is there yet. To show this pristine place, I photograph Amazônia alive, not the dead Amazônia."[15]

Forests in a warming world

As the climate catastrophe unfolds, forests are an important part of nature's carbon sequestration and storage and thus the planet-wide damage limitation. But will they still be able to play that role in a warmer world?

Atmospheric CO_2 levels were hardly dented by the disruptions of the COVID-19 pandemic and are now higher than—and changing more rapidly than—at any time in the last 23 million years. The annual average of the CO_2 levels recorded at Mauna Loa, Hawaii, passed 415 parts per million in 2022, compared to around 315 parts per million when measurements started there in 1958 and a pre-industrial level of 280 parts per million determined from ice cores. As a result, temperatures are increasing and are already locked in for a rise of more than 1°C above pre-industrial conditions.

The temperature increase affects different geographic areas and ecosystems in different ways. As has been widely reported, the Arctic is already experiencing some of the most dramatic temperature increases and ecological impacts, with the potential for feedback reinforcements contributing to a catastrophic outcome. Coral reefs and other marine ecosystems are already at risk of disappearing due to the combination of temperature effects, ocean acidification, and sea-level rise.

Tropical forests, in contrast, are more often discussed under the assumption that we must save them from our own economic greed so that they can save us by taking up some more of the carbon dioxide we release. Naïvely, some might assume that with their adaptation to heat and appetite for carbon dioxide, tropical forests might thrive even better in a global greenhouse and thus contribute to a dampening feedback effect. A comprehensive study has shown, however, that there are limits to the ability of tropical forests to help us out.

Just how the tropical forests will fare in a warming world, leaving aside human tendencies to decimate them for land-use change, has been a significant area of uncertainty in climate modeling so far. Studies of forests' resilience to climate change were typically based on extrapolations of their responses to short-term changes, such as from one average year to the next one if it turns out to be hotter or drier.

Such studies have often pointed to a sensitivity of tropical forests to climate warming. Their weak spot appeared to be the raising of the nighttime temperatures, which has been linked to slower growth and reduced carbon stock. Tree mortality during excessive droughts has also been identified as a mechanism by which climate change can affect the ability of forests to provide carbon sequestration.

Martin Sullivan from the University of Leeds, United Kingdom, and colleagues from many institutions around the world chose a different approach, using spatial rather than temporal differences as a model for what will happen as climate change advances.[16] The researchers argue that the conventional studies using short-term changes as a basis to extrapolate to long-term effect, while providing valuable insights into future responses of forests, may miss the potential of forests as ecosystems to adapt over the longer term in different ways, such as by a changing community composition.

Therefore, the researchers measured biomass carbon and carbon flux in 590 globally distributed, permanent tropical forest plots distributed around the tropics. For each plot, the authors determined the biomass present above ground to estimate the amount of carbon bound in the plants (the carbon stock). To gain information into how much carbon is sequestered over time, they calculated how fast the system gains carbon and how long it can retain that carbon (calculated as the ratio of living carbon stocks to carbon gains, in years).

Comparing between continents, the researchers found lower carbon stocks in South America and attributed the higher carbon stocks to faster gains in Asia and longer residence times in Africa. These differences, which may be related to the evolutionary history of the forest ecosystems on different continents, had to be considered in the analyses.

Based on their modeling of the data derived from climate differences between otherwise comparable plots, the authors concluded that the carbon stocks were most strongly linked to the maximum temperature, with the second-strongest influence coming from precipitation. That is, the amount of carbon stored benefits from lower temperature maxima in the warmest month and from more rainfall in the driest month. In contrast to the previous studies based on short-term effects, they found no significant effect of temperature minima.

Further analyses revealed that the adverse effect of higher temperature maxima is mainly due to a reduction of carbon uptake over time. This can be explained with the mechanisms of closing

pores at higher temperatures to avoid water loss, as this also reduces the access of carbon dioxide. By contrast, the effect of precipitation in drier months was mediated via the residence time.

Overall, Sullivan and colleagues conclude from their analyses
that tropical forests have so far kept their ability to help maintain
high carbon stock and have some capacity left to cope with further
temperature increases—within limits.

They also identify a tipping point, however, beyond which tropical forests will no longer be able to cope and could start to lose
their ability to sequester carbon. The thermal threshold, above
which the authors expect long-term losses of carbon stock, is at a
maximum temperature of 32.2°C. If and when global temperatures pass the 2°C warming threshold, more than 70% of current
tropical forests will be exposed to temperatures above this value.

South American forests in particular, the authors warn, are at
risk of passing this threshold and losing the ability to play a positive role in carbon sequestration. Could South America's tropical
forests shift to higher altitudes in the Andes? "Mountain forests
will be important stores of carbon in a warmer world," Sullivan
explains, "but the area they cover is less than lowland forests so
they can't compensate, and these areas also have experienced disproportionately high deforestation."[17]

To maintain the beneficial role of tropical forests in mitigating
climate change, therefore, we must not only save these forests from
further destruction but also keep temperature increases below the
2°C threshold. Meanwhile, forests have already changed dramatically and continue to change due to anthropogenic disturbances
including climate change. Only very few forests remain in a truly
natural, undisturbed state.

One important factor is the continuing fragmentation of forests
due to land-use change and construction of infrastructure such as
roads. This creates additional edges where ecosystems become

vulnerable to external disturbances, including extreme weather events, fire risks, and resource extraction, that may reduce them below critical size thresholds for some animal species, as will be discussed further on in this chapter.

In a study covering 18 years of forest fragmentation in the tropics, Matthew Hansen from the University of Maryland and colleagues found that the likelihood of forest loss increases for smaller forest fragments, even within protected areas.[18] This appears plausible as smaller fragments are also more accessible for disruptive influences like illegal resource extraction. The authors conclude that their findings "illustrate the need for rigorous land use planning, management, and enforcement in maintaining large tropical forest fragments and restoring regions of advanced fragmentation."

In a meta-analysis of reported changes in forest ecosystems, Nate McDowell from the Pacific Northwest National Laboratory in Richland, Washington, and colleagues showed that forests are already changing in response to human and environmental disturbances and trends.[19] The authors identified a global trend toward forests composed of younger and shorter trees due to the higher mortality and turnover caused by disruptions.

Even where reforestation programs have attempted to replace lost forests, they are still far from matching the biomass captured in and the ecosystem services provided by undisturbed primeval forests.

"This trend is likely to continue with climate warming," McDowell said. "A future planet with fewer large, old forests will be very different than what we have grown accustomed to. Older forests often host much higher biodiversity than young forests and they store more carbon than young forests."[20]

While there has been some evidence of increased growth due to CO_2 fertilization in the research analyzed, the authors conclude that this beneficial effect is limited by the availability of nutrients and water. Droughts can increase mortality directly or through making trees more susceptible to insect pests and diseases. This scenario has already been observed in Germany.

Germany has a very deep cultural connection with its forests, even though not very many of them survive. The existing forest areas are mostly exploited commercially for wood production. Only 5% of forest land is left alone and allowed to grow in a quasi-natural way, and even these forests are far removed from the deep dark woods of the brothers Grimm that have permeated the national psyche. And yet, when swathes of dying trees can be seen from the roads and railways, the country starts paying attention to the environmental problems that are killing the forests.

This happened in the 1980s, when the original *Waldsterben* phenomenon, caused by acid rain and pollution from coal-fired power stations, led to a surge of environmentalism and the rise of the Green Party. It is happening again, as forested areas like the Harz, which sits at the geographic center of gravity of the country, have seen massive die-offs since the drought of 2018.

This time, the air is much cleaner than it was in the 1980s. Instead, suspicion falls on climate change, which drives both the high temperatures and the more frequent droughts that weaken the forests. Combine that with the fact that entire forests are monocultures of economically lucrative species like spruce, which are sensitive to drought because their roots don't go very deep, and you get an ideal breeding ground for pests like the notorious European spruce bark beetle (*Ips typographus*), which is known for destroying spruce plantations at a rate that endangers their commercial viability.

In 2019, Angela Merkel's government organized a crisis summit and promised half a billion euros to address the problem, but the debate about what to do with that money very soon became heated. Environmental organizations wanted more natural and more diverse forests, while many commercial operators feared for their viability and just wanted to kill all bark beetles and return to their business as usual. The massive use of insecticides like sulfuryl fluoride, for instance, to fumigate wood from forests affected by bark beetles, and to make the wood safe for export, stirred some controversy, especially because the substance is a potent greenhouse gas.

Some enterprising scientists started thinking about replacing the spruce monocultures with new species from warmer regions, to match the climate that Germany will probably have by the middle of the century.

Jürgen Bauhus's group at the University of Freiburg, for instance, studies the resilience of 14 different tree species that are not widely grown in Germany but could become useful as climate change takes hold. Among the more successful species in his experiment is the oak species *Quercus cerris*, which hails from southeastern Europe and has traditionally been grown for the use of its bark by tanners. In 2019, Bauhus told a radio program on the ARD network that he sees oaks in general as potential winners in the upheavals caused by climate change. Commercial forestry operators are not rushing to bring out the acorns though, as the oak trees are more expensive for them.

In work conducted together with Lander Baeten from the University of Ghent, Belgium, and others, Bauhus studied the viability of various species for diverse and sustainable forestry in central Europe.[21] The study found that certain combinations of species can achieve both a high level of ecosystem functionality and a high productivity.

As climate zones move toward the poles, it appears inevitable that the species composition of forests will move with them. German forests may become more Mediterranean, and Scandinavian forests more central European. Reforestation programs, now widely supported as a measure to combat climate change, will have to take this into account and mind the diversity that will make plantations more resilient to pests.

Most importantly, however, forests can only help us to mitigate climate change if we stop decimating them further and don't push them over their tipping point.

Brazil's fragmented forests

Back in the 1970s, ecologists and conservationists were debating whether the insights from island biogeography (see chapter 6) also applied to small patches of conserved wilderness on dry land. Specifically, faced with the loss of wilderness and the urgent need to save some of it to avert mass extinctions, they asked whether it was better to preserve a given amount of land as a single large reserve or as several small reserves. The data available at the time was insufficient to give conclusive answers, so several researchers set out to address the issue.

Thomas Lovejoy, then at the World Wildlife Fund in the United States, devised a large-scale experiment in the Brazilian Amazon, which was officially launched in 1979 with the backing of Brazil's National Institute of Amazonian Research (INPA) at a site some 80 kilometers (50 mi.) north of the city of Manaus in the central Amazon. The study area comprises around 1,000 square kilometers (385 mi.²), which were at first studied in their natural state and then split up into fragments of different sizes, while the areas between them were converted to cattle ranches.

The fragments created for the project include two of 100 hectares (250 acres), four of 10 hectares (25 acres), and eleven of 1 hectare (2.5 acres) surface area. Larger areas of undisturbed forest nearby were also included in the study for control measurements.

More than four decades later, the project is now known as the Biological Dynamics of Forest Fragments Project (BDFFP), and it is run jointly by INPA and the Smithsonian Tropical Research Institute. It has become one of the largest biological experiments of all time and has generated vast amounts of results. It has produced over 700 scientific papers, more than 200 PhD and master's theses, educated hundreds of students, and hosted more than 1,000 interns.

In a comprehensive review of the lessons learned from the project, Bill Laurance from James Cook University in Cairns, Australia, and colleagues, including founder Thomas Lovejoy, analyzed not only the effects of fragmentation on the local flora and fauna but also how these effects interact with larger-scale disturbances, including those caused by climate change and extreme weather events.[22]

Major threats to smaller forest fragments tend to arise from the edge of the forest, where it is vulnerable to all kinds of disturbances. As the proportion of edge circumference to surface area becomes more unfavorable the smaller the size of the fragment, these edge effects are one of the main arguments in favor of larger, undisturbed protected areas.

In four decades worth of field studies at the BDFFP site, researchers have recorded a wide variety of edge effects. Most obviously, the open flank of a newly created clearing leaves the remaining forest more vulnerable to wind and weather. Wind shear and turbulence forces on the forest edge lead to a higher tree mortality. Desiccation caused by the inferior water retention capacity of the clearing adds to the damage. In comparison to trees, grasses and crops have smaller leaf surfaces and shallower roots, meaning that they can recycle a smaller fraction of the rainwater coming in (evapotranspiration). Research has shown that the drier air above a clearing can then suck the moisture out of the adjoining forest and cause desiccation penetrating hundreds of meters deep into the forest. Recent satellite measurements of moisture in the canopy even indicate that the effects can still be felt 2.7 kilometers (1.7 mi.) away from the nearest forest edge.

Over the years of the observations, these factors have led to changes in the species composition of the areas near the edges. Long-lived, slow-growing trees become less prominent, and fast-growing plants like lianas take over. To a certain extent, this change in composition can help to seal the surface of the forest fragment and thereby protect it from further damage from the microclimate. Wind tunnel measurements suggest, however, that even the mature edge remains vulnerable to storm damage.

These changes also affect the animals of the forest in various ways. Some are threatened, others benefit from the new structures. Ricardo Rocha from the University of Lisbon, Portugal, has studied these effects for the numerous species of bats prevalent in the forests.[23] He found, for instance, that predatory bats tend to decline near the edges, while some species of frugivorous bats benefit from the disturbance.

If several edges are nearby, the disturbances may add up, which explains why fragments of less than 10 hectares (25 acres) and those with very irregular shapes suffer significant alterations. Even relatively narrow cuttings for roads with no more than 30 meters (100 ft.) clearing width can have severe edge effects on both sides.

While the continuous presence of researchers at the BDFFP site has protected the forest fragments from many of the human-made disturbances seen elsewhere in the tropics, including fires, illegal logging, and hunting, edges and roads are generally known to expose forests to these threats. For instance, the observed desiccation of forests near the edge means that human-made fires can spread more readily in the forests. Roads cutting through forests, such as those built to supply construction sites of new hydroelectric dams, are often observed to become nuclei of deforestation, as hunters and loggers like to use them for illegal activities.

Other reasons small fragments support less biodiversity than larger ones include the sample effect as well as the critical habitat size required by some species. The sample effect is based on the observation that many of the species in the tropical forest are dispersed quite thinly, such that a smaller fragment, when it becomes isolated from the bulk forest, may not contain a sufficiently large population of the species to make it viable in the long term. This, along with other small and random effects, explains why the composition of some of the very similar fragments in the study diverges significantly over the years.

Critical size effects are already known from island biogeography, with the key difference here being that the clearings between the forest fragments may in certain circumstances represent a less efficient barrier than the water between islands. Therefore, the study of the fragments also had to consider the land in between, called the matrix.

The land between the forest fragments of the BDFFP site was initially used for cattle and thus kept clear of forest regrowth. However, after 1988, the loss of government incentives and the generally poor productivity of the ranches led to their demise. Since then, some regrowth in the matrix has softened the isolation of the forest fragments.

Small differences in the starting conditions of the land have sent different parts of it on different itineraries. Some parts of the matrix land were initially cleared and burned, while others, due to wet weather conditions at the time scheduled for the clearance, were cleared without the use of fire.

Initial regrowth on the land was dominated by *Vismia* shrubs in the areas that had been burned and by *Cecropia* tree species in the other parts. After the demise of the cattle farms, the *Cecropia*-dominated land developed secondary, species-rich forests with tall trees of up to 20 meters (65 ft.) in height, while the *Vismia*-dominated land recovered much more slowly.

As part of the fragmentation studies, some additional clearing was carried out to maintain the isolation of the fragments. Typically, stripes of 100 meters (330 ft.) in width were cleared around each fragment to afford well-defined isolation conditions.

Studies conducted at the BDFFP site show that the nature of the matrix is a key factor in determining how strongly the ecosystems in the fragments are disturbed. If the area is kept clear for agriculture, for instance, it can be an unsurmountable barrier for many animal species. Experiments with understory insectivorous birds have shown that some of these species can be forced to cross a road clearing some 50 meters (165 ft.) wide but not a pasture 250 meters (820 ft.) wide. Even modest amounts of regrowth can reduce that barrier and soften the edge effects in the forest fragments.

As these differences affect the movement of pollinators, the composition of the matrix can influence the plant communities in the fragments. Thus, the species composition differs markedly in fragments surrounded by *Vismia*-dominated (post-burning) matrix and those surrounded by *Cecropia* regrowth.

Beyond the immediate surroundings of forest fragments, regional and global effects can exacerbate the problems that forest ecosystems face. As mentioned, the desiccation of the air over clearings can threaten the water supply of the forest. On the regional level, some researchers have warned that the entire hydrological system could be threatened by forest fragmentation, which would have severe economic consequences reaching as far as Argentina cattle ranches, which depend on the Amazon's water pump as a vital ecosystem service. Moreover, the desiccation makes forests more vulnerable to fires, which in turn can produce a feedback loop by interrupting cloud formation, leading to region-wide drought conditions.

How the shrinking forests will interact with global change can only be predicted by large-scale modeling, which so far has yielded contradictory results. Some studies predicted hotter, drier conditions with catastrophic forest losses, while others were less pessimistic. One crucial factor to be considered is to what extent the higher CO_2 concentrations in the atmosphere will promote plant growth and facilitate water retention (as plants can reduce the time they have to open their stomata for CO_2 fixation at the risk of water loss).

Extreme climatic events are another crucial factor. They are known to become more frequent due to climate change but are still difficult to classify and quantify. One key advantage of the long time frame covered by the BDFFP observations is that it includes both rare events (such as the 1997–1998 El Niño, which severely impacted the Amazon) along with a reliable baseline of normal weather variation.

Overall, the BDFFP observations, running for over four decades now, have created a unique opportunity to better understand the effects of deforestation from the local to the regional level and to better understand deforestation's interaction with global change. Addressing the initial question about the size of reserves, Laurance and colleagues conclude in their review that reserves should be large and numerous. A good size likely to maintain natural ecological processes and balances would be 10,000 square kilometers (3,900 mi.²), or an undisturbed square plot of 100 kilometers (62 mi.) side length. However, political developments that occurred in Brazil under Jair Bolsonaro's presidency suggested that the number of such functioning reserves were more likely to shrink rather than to expand.

Over the last few decades, Brazil has been at the forefront of deforestation and the fight to reduce it and save the tropical forests. Illegal land clearing that was later regularized with a little financial encouragement paid from the gains made on the new agricultural land was a long-running problem in Brazil throughout the twentieth century, both under the military dictatorship of 1964–1985 and then under President José Sarney.

Brazil led the world in deforestation statistics, and saving the Amazon rainforest became the battle cry of the environmental movement. Good intentions were shown at the 1992 Earth Summit in Rio, and some nature reserves were created, but as an investigation from the British Sunday paper the *Observer* showed, some of them only exist on maps, and they have failed to deter the land grabbers.

Only after the turn of the millennium, under President Luiz Inácio Lula da Silva (2003–2011), did state controls manage to slow down the march of forest destruction. Using satellite technology and ground level control, the government authorities managed to slow down the decline of the Amazon significantly. Brazil thus earned respect in international environmental efforts such as REDD and was held up as an example to other countries that have shrinking tropical forests, such as Indonesia.

However, under Lula da Silva's successor, Dilma Rousseff (2011–2016), the agriculture lobbyists, known as ruralists, and their calls for more land to be cleared to produce soy and beef became more influential again. In 2012, Rousseff relaxed the main legislation against forest clearing, and observers have estimated that deforestation rates have increased by 75% since then. At a rate of 8,000 square kilometers (3,100 mi.²) per year, deforestation then proceeded at more than twice the rate that Brazil had pledged to achieve by 2020 in international climate talks.

After Dilma Rousseff was impeached in August 2016, former vice president Michel Temer became president, followed by Jair Bolsonaro in January 2019—neither known for any inclinations to look after the environment. At the beginning of 2023, Silva returned as president and started to resume his efforts to protect the Amazon.

In addition to rubber, soy, and beef, another factor (this one relatively new) in the Brazilian equation is palm oil, which has previously been linked to deforestation in Indonesia but not in Brazil. Theoretically, however, Brazil has the world's largest reserve of land that could be used for palm-oil production, according to an analysis by Johannes Pirker and colleagues at the International Institute for Applied Systems Analysis, Laxenburg, Austria.[24] With the weakening legal protection of the Amazon forests, there is a risk that Brazil will further expand in the lucrative palm-oil sector at the price of accelerating deforestation.

Sustainable palm-oil protection could be achieved on lands that are already in agricultural use, if farmers make the switch from other crops or cattle ranching to growing oil palm, as Rhett Butler from the environmental website Mongabay has suggested. This could make Brazil independent of palm-oil imports and improve the economic viability of some agricultural regions.

However, many fear that the riches promised by the global demand for palm oil may lead to a new acceleration of forest losses, as they already did in Indonesia.

How can we save forest biodiversity?

Deforestation has been recognized as a global problem threatening wildlife habitat as well as multiple ecosystem services that benefit human economic activity on a massive scale. The need to limit carbon emissions and climate change, in particular, is critically dependent on the contribution of forests.

In 2008, the United Nations Collaborative Programme on Reducing Emissions from Deforestation and Forest Degradation in Developing Countries (UN-REDD) was launched to encourage forest conservation in a framework that also allows sustainable development. In addition, several countries signed up to the REDD+ program, which aims to incentivize forest conservation in developing countries, such as Indonesia and Ecuador, with funds from major corporations and wealthy nations like Norway. However, the Center for International Forestry Research reported in 2016, after a meeting of experts from 16 countries that have embarked on the REDD+ process, most of the participants hadn't advanced to the stage of actual fund transfers yet. Short-termism in politics was cited as a main reason why the pursuit of this long-term project is struggling in many of the countries.

While organizations and governments are looking for ways to halt deforestation, the bad news has kept coming in, showing the demise of forests around the globe, ever more efficiently monitored with improving satellite technology.

Although the continuing losses of forest area as observed by satellites are an important concern, the mere surface area alone doesn't tell the whole story. We have discussed above that fragmentation is an important issue. And when it comes to the benefits that forests provide, the diversity of plant species populating that area is just as important, as a global study has demonstrated.

Many regional studies have suggested that the productivity of forest ecosystems increases with increasing tree biodiversity. Jingjing Liang, a forest ecologist at West Virginia University in Morgantown, had noticed this trend in his field studies and wondered if this was generally true, as has previously been shown for herbaceous plants but not for forests.

He asked colleagues around the world for their observations and compiled datasets on biodiversity and productivity from over 700,000 plots in 44 countries. More than 30 million trees of over 8,700 species were measured at least twice to assess their timber production over time, which served as a simple and economically relevant measure of ecosystem productivity.

The overall results show a consistent correlation between diversity and productivity.[25] This finding makes sense in broad ecological terms, as a diverse plant community is better equipped to make the most of all the resources available in each environment, to cope with short-term variation, and to withstand pest species.

The graph showing the productivity as a function of diversity displays a downward curvature with a gradient that becomes steeper at low diversity. This means that a highly diverse community will be fairly resilient to the effects of losing one or two species, but as diversity declines, the loss in productivity per species lost becomes more pronounced.

The impact of species loss also varies geographically. The authors found that a given percentage of species loss would lead to a greater relative decline in productivity in the boreal forests of the Northern Hemisphere—although in terms of absolute productivity numbers, the effect would be stronger in more productive regions such as the Amazon.

The authors concluded that the current species loss in forest ecosystems around the world could dramatically reduce forest productivity as we approach the sharp fall shown in their graphs. Beyond timber production, which they used as the output parameter of the study, the same relation is likely to translate to other services the forests supply, including carbon sequestration, soil stabilization, water purification, habitat for other species, and regional rainfall regulation.

Without even taking these other services into account, based on the timber production alone, the authors calculated that the economic value provided by the remaining forest diversity amounts to $166 billion to $490 billion per year. This figure, although it represents only a fraction of the total benefits of tree biodiversity, already exceeds the cost of effectively conserving all terrestrial ecosystems on the planet, which in 2012 was estimated at $76.1 billion per year.[26]

Thus, this analysis suggests that the age-old perception of a conflict between ecology and economy is a red herring. Globally, and in the long term, conserving forest biodiversity will be crucial to saving both wildlife and the economy.

If biodiversity is crucial for productivity and ecosystem services, then conservation efforts should, and often do already, focus on this aspect. However, biodiversity is not only found in the places that are famous for it, like tropical rainforests, and it isn't always as well protected as it should be.

The seasonally dry tropical forests of the Americas, for instance, are much less appreciated for their biodiversity than the rainforests. By compiling an inventory of plant species in these biotopes, the Latin American Seasonally Dry Tropical Forest Floristic Network (DRYFLOR, http://www.dryflor.info/) located the species-rich areas that should be protected.[27]

The study looked at forest areas with a closed canopy receiving less than 1,800 millimeters (71 in.) of rain per year and experiencing a drier period of three to six months per year. These areas, spread from Mexico to Brazil, hosted pre-Columbian civilizations and are still attracting high human population densities with their fertile soils and moderate climatic conditions. Therefore, many of these biotopes are already highly degraded.

Analyzing the species distribution in these biotopes across Latin America and the Caribbean, the DRYFLOR researchers found that they divide into 12 floristic groups, each of which has its characteristic species repertoire. The groups can be organized into two clusters with closer connections; namely, a northern cluster, from Mexico down to Colombia and Venezuela, and a southern cluster.

This division, which has also been suggested by previous studies, mainly reflects the fact that the Amazon rainforest presents an effective barrier for virtually all species of the dry forests.

The researchers found that species richness depends on the size of the area, with larger areas predictably hosting larger numbers of species, as was found by the fragmentation studies discussed earlier in this chapter. By contrast, the effect of latitude is less pronounced and proximity to the equator does not enhance species richness in these biotopes. The authors conclude that, if anything, the biodiversity of seasonally dry forests peaks around latitude 20°.

A key finding was the diversity between these forests. The researchers found that no species is shared between all 12 groups. Only a few species occur across 10 or 11 groups, and these are typically generalists that are also found in other kinds of environments, including rainforests.

In terms of conservation concerns, this means that each of the 12 groups has its unique brand of biodiversity worthy of protection. Nevertheless, two of the groups occurring in the Andes have no protected area at all, and overall only 14% of the area of inter-Andean dry forest falls within protected areas. For the entire set of sites in the DRYFLOR database, the protection rate is also 14%. This figure falls short of the UN's Aichi biodiversity target of 17%.

The authors concluded that "it is clear that current levels of protection for neotropical dry forest are woefully inadequate." They call for conservation policy to prioritize the still unprotected sites in the Andes, while noting that the high population pressures on these areas will make it necessary to consider the social dimension and offer opportunities and incentives to local communities and landowners. They also suggested that, based on the DRYFLOR database, some of the plant species involved could be protected using IUCN Red List criteria, which might help to boost awareness of the threats these species are facing.

Forests are a key element of the climate crisis. They make a vital contribution to the recycling of carbon dioxide from the atmosphere, but they are threatened by rapid changes in climatic conditions as they cannot migrate on compatible timescales. In an analysis based on fossil evidence of plant migration under climate change over the last 21,000 years, David Nogués-Bravo from the Natural History Museum of Denmark in Copenhagen and colleagues found that significant range shifts of many plant species would be necessary to cope with climate change by 2050, although extinctions caused by climate alone would be rare.[28]

In this context, the species and genetic diversity of forests is crucial not just to their productivity, as outlined above, but also to their survival, as diversity is required to be able to adapt to change.

Our civilization has so far done its utmost to diminish forests both in size and in diversity. Where native growth is replaced by cultivated forests for industrial purposes, from timber to palm oil, these are typically grown under conditions close to monoculture.

Clearly a different approach to forest management is needed if we want to keep the benefits that forests are providing. Inspiration regarding sustainable cohabitation between humans and forests could come from the Indigenous populations who have lived in tropical forests for tens of thousands of years.

As discussed earlier in this chapter, there is ongoing research and debate over the extent to which tropical forests that were hitherto considered pristine have in fact been managed and exploited by forest-dwelling tribes over millennia. It also emerged that ancient cities in forests, including the Mayan cities and Angkor Wat in Cambodia, fed much larger populations than had previously been considered possible.

The last surviving groups of tropical forest tribes are currently facing the threat of extinction due to the ongoing destruction of their forests. Linking their plight to the need to conserve forests to combat climate change, the British economist Nicholas Stern (author of the eponymous climate report to the UK government in 2006) suggested that protection of Indigenous land rights would simultaneously create the maximal environmental benefits. Stern highlighted this connection at a recent World Resources Institute event for the launch of a report quantifying the amount of CO_2 emissions that could be avoided annually just by securing Indigenous land rights in Bolivia, Brazil, and Colombia. The report, titled "Climate Benefits, Tenure Costs: The Economic Case for Securing Indigenous Land Rights in the Amazon," found that forests with legally recognized Indigenous land rights have suffered much smaller losses than those without such rights.[29] In economic terms, the report concludes that "the modest investments needed to secure land rights for indigenous communities will generate billions in returns—economically, socially and environmentally—for local communities and the world's changing climate."

Meanwhile, Indigenous communities across the Americas are at the forefront of protests against land-use change and infrastructure projects, like the Dakota Access Pipeline in North Dakota—often they are the most vocal critics of fossil-fuel industries. Maybe it is time that our allegedly advanced civilization stops inflicting its business model on Indigenous forest-dwelling populations and asks them for advice instead. They have demonstrated, after all, there is an alternative to our destructive ways, and that humans can live in, with, and from forests.

CHAPTER 5

This Time, the Asteroid Is Us

We know of five global mass extinctions from the geological record. The most recent one, around 66 million years ago, is widely known for wiping out the dinosaurs (apart from those that evolved into birds) and thereby favoring the rise of mammals and birds, which expanded to fill the ecological niches left deserted and to grow back a new biodiversity. That extinction has been studied in detail. We now can say with a fair bit of confidence that it was caused by the climate impact of the 10-kilometer-wide (6 mi.) asteroid that produced the Chicxulub crater in the Gulf of Mexico.

We are now experiencing the beginnings of what is likely to become the sixth mass extinction. This time, the asteroid is us. Humans evolved in Africa in balance with their ecosystem. When they started to invent sophisticated long-range hunting weapons and then expanded out of Africa and around the world, they became an unprecedented and unnatural kind of predator and caused regional extinctions wherever they went.

Now, by killing the wrong kinds of animals, destroying habitat, and forcing climate change, we are causing a global loss of biodiversity on a scale that hasn't been seen since the dinosaurs disappeared. We no longer fit in as part of a natural ecosystem but are destroying the natural systems that keep us alive. Ironically, this happens just in the moment when we are beginning to learn from ecology and other sciences how these natural systems work and why they are so important.

Hunting wildlife to extinction

Before humans started to spread around the world, there were large animals on all continents. Now most of these have disappeared, and we have replaced their biomass, but not their ecological function, with cattle and other domestic breeds. A sizable fraction of the Pleistocene megafauna has only survived in Africa, where animals witnessed the rise of human hunters and learned to fear them, and in the oceans, which remained quite inaccessible until a few centuries ago (see chapter 6).

Only within this century have we learned that by eradicating large animals like rhinoceroses we are losing more than just natural diversity and spectacular sights for safari tourists to photograph. The large animals have been making a disproportionate contribution to ecosystem services, like nutrient distribution against the hydrological cycle (see chapter 9). Their disappearance may thus not only affect other animals and humans, but it could destroy forests and thus habitat and other ecosystem services.

In a first large-scale systematic study of hunting-induced declines of animal populations, Ana Benítez-López at Radboud University in the Netherlands and colleagues found that in areas within easy reach of access points such as roads, mammal populations are reduced to 17% of their abundance in hunted areas compared to unhunted areas, while bird populations are reduced to 42%.[1]

In a follow-up to this study focusing on mammals in tropical forests around the world, Benítez-López and colleagues concluded that more than half of tropical forests are under severe hunting pressure.[2] "Even forests that are considered intact according to satellite images—in which there is no visible deforestation or logging—could be partially defaunated," Benítez-López explained.[3]

According to models, based on more than 160 ground-based studies from a wide range of locations and more than 3,200 mammal abundance estimates, the researchers predict that 52% of intact forests and 62% of wilderness areas in the tropics are partially devoid of large mammals, and even in protected areas, 20% of the surface area is suffering from hunting pressure.

Some regions are particularly affected. "We found the biggest declines in Western Africa, with more than 70% of population reduction. Our calculations show that even in protected areas mammal populations could be under hunting pressure, particularly in Western and Central-Africa, and South-East Asia," Benítez-López said. The situation is relatively less problematic in the Guianas and in the Brazilian Amazon.

Access to the forest areas is a key parameter driving the hunting pressure. As infrastructure programs further fragment the forests and facilitate access, the problem is bound to grow. Benítez-López and colleagues warned of the danger of "intact but empty forests." This could impact the net carbon storage capacity of forests, as the lack of seed-dispersing and herbivorous animals will favor fast-growing plants such as lianas over trees that are more useful in terms of ecosystem services.

Thus, the very existence of the remaining tropical forests could be further diminished by hunting, in addition to, and in synergy with, other human influences such as land conversion for agriculture, infrastructure building, or mining. The authors concluded that "retaining the integrity of intact tropical forests will not be possible if global and national environmental strategies do not address ongoing hunting practices."[4]

Although much of the terrestrial megafauna has already disappeared and average body size of animals has consequently shrunk, hunting continues to shift the size distribution of animals in favor of smaller animals that are less attractive targets for hunters.

A study by Robert Cooke from the University of Southampton, United Kingdom, and colleagues predicted that this trend is going to continue.[5] Simply by checking IUCN Red List data against attributes like body size and ecological roles, the researchers were able to predict that within the next 100 years, average body size in birds and mammals will shrink further due to the expected extinction of more than 1,000 species. Across the parameters studied, the surviving species will occupy a narrower range of ecological roles, with a higher proportion of mammalian and bird species tending "towards small, fast-lived, highly fecund, insect-eating generalists," according to the authors.

Even the most iconic animal species enjoying global conservation support, such as elephants, aren't safe from poaching and changes in regional politics. In May 2019, the new government of Botswana shocked conservationists by lifting a ban on elephant hunting. While it is true that elephants have been thriving in Botswana, they are still declining in Africa overall, with illegal hunting being the major cause of their demise.

Compared to the terrestrial megafauna, inhabitants of the oceans have remained relatively unscathed due to the inaccessibility of their habitat. We will come back to the very different situation of the marine fauna and megafauna in chapter 6.

One might imagine that birds, mostly equipped with the ability to fly and in some cases also with remarkable intelligence, would be in a privileged position to escape the efforts of human hunters, but then again, long migrations along known routes with necessary touchdowns make many species vulnerable to hunting.

A cause célèbre of a species that is literally being hunted to extinction is the ortolan bunting (*Emberiza hortulana*), a small migratory songbird that passes across southern France on its migration toward wintering grounds in Africa. Although the European Union (EU) has listed it for protection under the Birds Directive since 1979 and France passed legislation to protect it in 1999, the tradition of capturing, fattening, and eating the birds continues unabated, with an estimated 30,000 killed in France every year.

Local hunters, chefs, and politicians have defended the practice as a cultural tradition, arguing that the birds caught in France come from populations robust enough to sustain these losses. To establish the real impact of hunting on this declining species, Frédéric Jiguet from the Muséum national d'Histoire naturelle in Paris, France, and colleagues conducted a comprehensive and detailed study of the migration patterns of the species, combining modern tracking methods with genetics and isotope measurements.[6]

Based on the findings and on viability modeling, Jiguet and colleagues concluded that the ortolan buntings caught in France are from the declining and fragmented northern and western populations of the species, and that the current level of harvesting, which locals want to establish legally, is far from sustainable. According to the authors, the results provide "sufficient scientific evidence for justifying the ban on ortolan harvesting in France." Now all that is needed is the political will to enforce a ban that has been in place for 20 years and has been violated hundreds of thousands of times.

Similar problems with traditional bird hunting practices continuing in violation of protection laws are found all around the Mediterranean. Problems are heightened on islands such as Malta that serve as stopovers on bird migration routes and host unrepentant hunters. Malta in particular is a crucial crossroads of the migration routes of around 170 bird species from 48 countries, according to a report by BirdLife Malta.

Since joining the EU in 2004, Malta has gained a track record of claiming cultural exceptions (known as derogations) to EU conservation rules and exploiting EU leniency. In June 2018, the European Court of Justice ruled against Malta on the issue of finch trapping, for which Malta had claimed a derogation.

But even though Malta respected the court order and did not issue a derogation for finch trapping in 2018, other groups of birds are still being trapped on the island, and larger birds, such as storks, risk being shot dead as soon as they arrive there, as Jason Greggs reported on the environmental website Mongabay in January 2019.[7]

After Malta, Cyprus is known as the second-most deadly stopover island for birds migrating across the Mediterranean. An investigation by the United Kingdom's Royal Society for the Protection of Birds found that trappers on Cyprus killed 2.3 million migrating birds in the fall of 2016. As a large part of the illegal hunting goes on in two areas under control of the British military bases located there, the British authorities have reportedly started a clampdown on illegal bird hunting, using technology including drones, and they are beginning to see a successful reduction in the estimated number of killings.

While the Mediterranean tradition is to kill birds for food, birds in the tropics are at risk of being trapped for the pet trade. Historically, this has affected various species of parrots, but it is now expanding explosively, even to species completely unsuitable for the use as pets, according to Cagan Sekercioglu from the University of Utah in Salt Lake City.

Sekercioglu has been involved in a study to determine the impact of the pet trade on birds in Sumatra, Indonesia.[8] Based on field surveys, interviews with trappers, and analyses of market prices, the researchers found that the remaining forests offer little protection to the wild birds and that their decline is tightly coupled to market demand.

Whether birds are hunted for food or for amusement, the markets drive unsustainable hunting around the globe. With biodiversity already on the brink of a sixth mass extinction, it is high time to put a stop to these practices.

Shopping with hunter-gatherers

Modern *Homo sapiens* individuals, across much of their temperate habitat, are often observed burning hydrocarbons to move several tons of steel for a kilometer or two, just to acquire a few kilograms of food. Their foraging depends on fellow humans elsewhere who have spent even more energy producing the food stuff and transporting it to the shops. As well, food must be heated or otherwise treated to make it more digestible and nutritious.

The overall energy efficiency of the effort spent to feed a few kilocalories into a human metabolism may appear questionable to an alien observer. In the face of a mounting climate catastrophe, Earthlings can no longer deny that all this is bad for their planet. Convenience and time efficiency are the arguments that win over energy efficiency and sustainability. New research in anthropology shows that this trade-off goes back a long way, to the point in our evolutionary past when hominids took up hunting and gathering.

Looking back from a point in human history some 12,000 years after the origins of agriculture, we tend to regard the hunter-gatherer lifestyle as a human baseline from which agriculture and further technological innovations departed. But seeing that our closest relatives among the apes, the chimpanzees and bonobos, don't do anything comparable, it is clear that this lifestyle also evolved at some point (maybe 2.5 million years ago) and shaped the further evolution of our species at a stage when other directions might have been possible.

Thomas Kraft from the University of California, Santa Barbara, and colleagues analyzed the energetics of the hunter-gatherer subsistence strategy in comparison to that of other great apes.[9] The fundamental anthropological question motivating their research is, What aspect of the typically human way of life most facilitated the developments of characteristic human traits, such as extended childhood and child dependency, large brains, long life spans, and division of labor? This study has its roots in the long-running research into human energy expenditure from the group of Herman Pontzer at Duke University in North Carolina.

The researchers compared subsistence costs (i.e., energy and time invested in relation to the energy obtained) among wild orangutans, gorillas, and chimpanzees and compared them to detailed data on the food acquisition of the Hadza people, a group of hunter-gatherers in the northwest of Tanzania, and the Tsimane people, forager-horticulturalists in the Bolivian Amazon. They also assembled a global database with subsistence energetics information of contemporary hunter-gatherers and horticulturalists.

Naïvely, one could have imagined that, to free energy for child rearing, for instance, humans might have evolved energy-saving measures. The sloth is evolution's prime example of how far energy saving can go, as we saw in chapter 2. Separate work on the energetics of walking in natural environments from the group of co-author Vivek Venkataraman from the University of Calgary, Canada, suggests that efficient ways of walking upright might have helped early hunters to be sophisticated energy savers.[10] Locomotion is typically the main energy cost of foraging, as Kraft and colleagues note.

Although many energy-saving methods evolved across the tree of life, "that doesn't mean natural selection always favors reduced

energy expenditure," Kraft explained. "In fact, tremendous variation exists in the 'tempo' of energetic strategies. A dramatic example is the difference between endothermic (warm-blooded) and
ectothermic (cold-blooded) animals. Warm-blooded animals tend
to use a lot more energy each day but are able to successfully channel that energy into activities that ultimately lead to successful
reproduction."[11]

In this spectrum of natural possibilities, Kraft and colleagues
found humans to be naturally generous with their energy budget.
They show that, compared with other great apes, humans readily
invest more energy to gain more food energy in a shorter time
span. For instance, making spears, assembling a group of hunters,
and going out to hunt a big beast requires an upfront investment of
effort, but, if successful, it may pay off with more energy yield than
just picking the low-hanging fruit you see around you.

With the development of efficient and ambitious hunting and
rapidly improving weaponry, humans catapulted themselves to
the top of the food chain, causing megafaunal extinctions, especially on the new continents they conquered after the migration
out of Africa, as we discussed earlier in this chapter.

The overall energy efficiency, in terms of calories gained per
calories invested is roughly comparable with the subsistence strategy of other great apes. The parameter that ultimately gives humans
an advantage is the time efficiency. With that extra effort invested,
they can produce more food in a given time, and more than they
need themselves. This makes it easier for them to provide for dependent children and older members of the group and creates free time
to pursue other things, such as social and cultural activities.

This mechanism had already been appreciated for the onset of
agriculture, where it resulted in the diversification and stratifica-
tion of society, as the excess food production enabled the new roles
of soldiers, priests, monarchs, and so on, and thus also the spread

of agriculture by conquest. Kraft and colleagues now show that a similar logic of freeing time and resources already applied more than two million years earlier, when humans first started hunting and gathering. In a sense it may also explain why we now make massive investments in technology to produce, distribute, and acquire fast food and convenient food, even if it isn't very good for us or the environment. Essentially, it looks like we evolved to get our energy fix quickly.

Investing effort and resources into future food provision inherently carries the risk of losing the investment. With the spread of agriculture and the storage of seed grains for the next season, these risks have grown. For some of these risks, like infestation with rodents, evolution found inventive solutions, like the auto-domestication of the cat (see chapter 7). Other risks, from extreme weather events to locust plagues, have remained and have produced deadly famines throughout human history. In today's industrialized and globalized food systems, disturbances to production and logistics can endanger the food security of millions, as the ripple effect of the war in Ukraine has demonstrated.

As hunter-gatherer groups are co-operating and collectively looking after infants, their overall success and energy balance also depends on the network structure and its information-processing capacity, parameters that can be studied in terms derived from computer science.

Marcus Hamilton from the University of Texas at San Antonio combined the energetics and computing perspectives in a recent analysis.[12] He noted that extrapolation from other primate species to humans would predict a likely band size of 32 individuals, but the empirical findings for hunter-gatherer groups suggest that the natural value is only half that, around 16 individuals.

Hamilton argued that humans could save themselves the expenses of larger group sizes as their unusually large brains, connected even to a small network, deliver the benefits much earlier. Thus, he found that a network of 16 human brains provided more computational power than any other social group of mammals. This combined brain power can help to predict and handle challenges from the environmental conditions as well as the necessity to find food.

At a higher level, human groups also exchange information as a metapopulation, giving access to computational power beyond anything available in the animal world. The metapopulation, Hamilton concluded, maximizes the benefits of networked intelligence while keeping the costs down. This balance produces the widely known small-world network, where everybody is connected to everybody else by no more than six degrees of separation.

After hunting and gathering, the next step up in time efficiency of food provision was the introduction of farming, with the domestication of crops and animals and the development of permanent settlements. This happened independently in a few regions around the world, the first of which was the Fertile Crescent in today's Middle East and Türkiye. This important transition, also known as the Neolithic Revolution, is well studied archaeologically and is now also accessible to molecular analyses of ancient genomes.

As the geographic origin of farming in the Fertile Crescent can be pinned down to a small region in Anatolia, a widely held assumption was that the population that first made the transition would be similarly narrow. However, a detailed study of ancient genomes presented a much more complex picture. Nina Marchi from the University of Bern, Switzerland, and colleagues sequenced the genomes of 15 individuals to high coverage (at least 10-fold) and then used the high-quality genomes for demographic modeling, in a combined approach they term demogenomic modeling.[13]

According to the models from Marchi and colleagues, the population shifts leading to the first farming population in Anatolia may have unfolded as follows. Some 25,000 years ago, a hunter-gatherer population based in the Middle East split into two parts when one moved into southeastern parts of Europe, while the other stayed in the area. During the Last Glacial Maximum, the inclement weather in Europe and permafrost reaching as far south as today's Hungary brought the European subpopulation close to extinction, producing bottleneck effects of low genetic diversity that remained visible in genetic studies over many generations.

Previous investigations had assumed that the low diversity in the descendants of Ice Age Europeans was due to a small group size. The new model, however, suggests that groups remained connected in an effective metapopulation, in line with Marcus Hamilton's findings discussed above. The population as a whole, however, struggled in colder phases and recovered when the temperatures rose.

When the glaciers finally receded and the European population could recover and expand, it reconnected with the population left behind in Anatolia, around 12,900 years ago. From this confluence of the western and eastern population emerged the first farmers, and their genetic legacy later spread across Europe.

In a different approach, studying many more individuals at a lower resolution, Eske Willerslev's group at the University of Copenhagen, Denmark, obtained results broadly in agreement with the model described above.[14]

Further analyses of the ancient genomes in conjunction with the archaeological information and the vast numbers of genomes we now have of modern humans may also reveal more about the physiological details of the hunter-gatherer lifestyle, such as which genetic traits were necessary and which were beneficial or otherwise in those conditions.

Over the last two million years, our species has pursued a relentless drive toward more time efficiency, from big-game hunting to agriculture, and through to today's industrial food production. As Kraft and colleagues concluded, "Energy gained from improvements to efficiency in human evolution were primarily channeled toward further ramping up foraging intensity rather than reducing the energetic costs of subsistence."[15] As the authors noted, this counterintuitive trend finds a parallel in the Jevons Paradox, a macroeconomic principle stating that the introduction of more efficient technologies leads to increased consumption rather than savings in human systems.

As we come to understand this development better, we may also reflect on the growing collateral damage that decimated our natural resources. Our perennial quest for faster food has caused or contributed to terrestrial megafaunal extinctions, deforestation, a near-death experience for several species of whales during the brief period of industrial whaling, and now the climate catastrophe. The hunter-gatherer genes that set us apart from the more relaxed lifestyle of the apes may have made us masters of the Earth, but we still have to learn to use our power sustainably.

Hard times for orangutans

On March 4, 2019, the state administrative court in Medan, the capital of the Indonesian state of North Sumatra, ruled against the Indonesian Forum for the Environment (WALHI), which had challenged the state government's plans to construct a dam for a 510 megawatt hydroelectric power plant, the Batang Toru dam. The $1.6 billion project, funded by the Bank of China as part of the country's gigantic "Belt and Road" infrastructure initiative, is now expected to be finished by 2026. The dam was originally scheduled to be completed by 2022 but was delayed during the pandemic.

WALHI and many others objected to the project because the dam and its access routes divide and shrink the Batang Toru forest, the only habitat of the rarest and most threatened of the three extant orangutan species, the Tapanuli orangutan (*Pongo tapanuliensis*), as well as other threatened megafaunal species, including the Sumatran tiger (*Panthera tigris sondaica*) and the sun bear (*Helarctos malayanus*). The decision marks a setback for the long-running fight against the reservoir and for efforts to improve conservation of megafauna in Indonesia more generally.

The orangutans today found on the islands of Borneo (split between Malaysia, Brunei, and Indonesia) and Sumatra (Indonesia) are the sole survivors of their genus (*Pongo*) and subfamily (Ponginae), whose range once included continental South Asia as well as these islands. Although still reminiscent of humans in features such as their hands and faces, their lineage left Africa before the diversification that led to humans. Thus, orangutans are our most distant relatives among the great apes.

Within the genus *Pongo*, the identification of species has been controversial and changeable. Linnaeus named *Pongo pygmaeus*, the Bornean orangutan, in 1760. René Primevère Lesson (1794–1849) described the Sumatran population as *Pongo abelii* in 1827. But both species were subsequently regarded as the same species, until genetic analyses proved that they were distinct.

In 2017, the groups of Erik Meijaard at the Australian National University in Canberra and Michael Krützen from the University of Zurich, Switzerland reported that the isolated population occupying the southernmost part of the range in Sumatra, south of Lake Toba, is in fact a third orangutan species.[16] Alerted by the distinct shapes of skulls, jaws, and teeth in an individual from the Batang Toru area, the researchers applied genomic analysis to establish this individual's relation to the two known species of orangutan. Surprisingly, the Tapanuli orangutan (*P. tapanuliensis*) turned out to be the most evolutionary distinct of the three species, with closer links to the Bornean species than to its neighbors on Sumatra. The two Sumatran species parted ways 3.38 million years ago, but the division between the two previously recognized species occurred much later, only around 674,000 years ago.

With an estimated population size of fewer than 800 individuals, the newly identified species swiftly became one of the most endangered primate species known. One might have hoped that the combination of extreme endangerment, remarkable evolutionary distinctiveness, and relative proximity to humans on the tree of life would help the cause of conservationists and make the threat to the Tapanuli orangutan a sufficiently strong argument in the fight to stop the construction of the hydroelectric dam set to divide the orangutan's habitat in two.

Indeed, the WALHI group and other local activists had been optimistic ahead of the court process. Apart from the obvious threat to the survival of a unique great ape species, the group has alleged in its lawsuit that there were flaws in the environmental impact permit and that the project violated risk-prevention rules, as the dam is to be built close to a known fault line where the risk of damage from earthquakes is high.

In addition, the environmental website Mongabay reported that Onrizal Onrizal, one of the scientists who worked on the environmental impact report until 2013, had his findings removed from the 2016 report. Onrizal alleges that he was not told of the publication of the report omitting his findings and that his signature on it was forged.

Onrizal's earlier version listed 23 species of conservation concern, while the final report has only 15. Missing from the current report are the Sumatran tiger, sun bear, and Sumatran lar gibbon, as well as the Tapanuli orangutan. Although the latter were in 2016 still believed to be Sumatran orangutans, that species is also critically endangered, and any measure impacting its survival is technically illegal under Indonesian law.

The company PT North Sumatra Hydro Energy, which is leading the development, has argued that the habitat loss is only a small fraction of the existing orangutan habitat and that the access routes will be led along the riverside such as to avoid further fragmenting the forests.

However, Gabriella Fredriksson, a wildlife biologist with the Sumatran Orangutan Conservation Programme, told Mongabay that there were so few Tapanuli orangutans that "this species can't afford even an offtake of a few individuals."[17]

Moreover, as Bill Laurance from James Cook University in Cairns, Australia, pointed out, new roads through forests always carry the risk of enabling illegal logging and hunting. "One would like to think that environmental safeguards for a venture as high profile as China's Belt & Road Initiative would be effective, or at least basically operable," Laurance commented after the court verdict. "But we're finding almost the exact opposite, as exemplified by the appalling decision by China and Indonesia to drive ahead with the Batang Toru dam project right in the heart of the tiny range of the world's rarest ape, the Tapanuli orangutan."[18]

If the Tapanuli orangutan is already teetering on the brink of extinction, the other two orangutan species are not much safer according to the "EDGE of existence" program (https://www.edgeofexistence.org/) of the Zoological Society of London, United Kingdom. The EDGE (Evolutionarily Distinct and Globally Endangered) program uses evolutionary distinctiveness in combination with global endangerment to suggest conservation priorities, and it has ranked both species jointly forty-third among the most threatened and distinctive in its list of mammals, which was prepared before the Tapanuli orangutan was described.

The critically endangered Sumatran orangutan is confined to the northern part of the island, where it survives in the remaining patches of rainforest, especially in Gunung Leuser National Park near Aceh, which is a protected area. Orangutan populations are notoriously difficult to quantify because of their elusive nature and arboreal lifestyle. In 2016, Serge Wich from Liverpool John Moores University, United Kingdom, and colleagues arrived at an estimate of 14,613 Sumatran orangutans based on transect surveys.[19] This estimate is more than double the value that was previously assumed. The authors explain the discrepancy with the broader scope of their survey, which also covered areas previously considered irrelevant, including those at higher altitude and those with degraded forests.

Although the population turns out to be more numerous than thought, the authors predict rapid decline in years to come. Habitat loss is the main threat here, as Sumatra suffers rapid deforestation, partly driven by palm-oil production. The connection between palm oil and the dangers to orangutans was highlighted in a short animation film produced by Greenpeace in 2018.[20] The clip became an internet sensation when a supermarket chain in the United Kingdom wanted to use it for its Christmas ads but saw it rejected by the vetting agency Clearcast for being political. Hunters working for the pet trade, who in order to abduct young apes typically kill their parents, are also contributing to the decline of the species.

With an estimated population of just over 100,000 individuals, Bornean orangutans are more numerous than the other two species but also in decline and classified as critically endangered by the IUCN. Maria Voigt from the Max Planck Institute for Evolutionary Anthropology in Leipzig, Germany, and colleagues compiled field surveys and conducted modeling studies. They concluded that the Bornean orangutan populations have shrunk by more than 100,000 since 1999.[21]

The authors reported that the steepest declines, not surprisingly, were found in areas where the habitat was destroyed. In terms of total reduction of population size, however, the losses within remaining habitat only partially affected by logging were more significant, suggesting that hunting and capture still pose a substantial threat. Based on habitat reduction alone, the researchers predicted losses of more than 45,000 individuals by 2050.

Although we are technically the "third chimpanzee," and thus more closely related to African apes than to Asian ones, orangutans are still close enough to appear almost human in many ways, from the shapes of their hands and faces to some aspects of their elaborate behavior.

In an experimental study on planned tool use in *Pongo abelii*, Isabelle Laumer from the University of Vienna, Austria, and colleagues found that the apes were able to resist instant gratification when choosing between a food item and a tool that would enable them access to a more desirable kind of food.[22] This finding aligns with many anecdotal observations of the intelligence and creativity of orangutans held in captivity, who find unexpected uses for items they encounter in the unnatural environment.

Humans and orangutans are further linked by many millennia of interactions that have occurred since humans first arrived in South Asia. Stephanie Spehar from the University of Wisconsin–Oshkosh and colleagues argued in a review article that the apes, which were once widespread across southern Asia, have long been shaped by interactions with humans.[23] As the situation stands now, however, humans will have to put in their best efforts to ensure that this interaction will not end up being fatal to the three surviving species of orangutan.

Last call to save the rhinos

Rhinoceroses roamed all over the European steppe as recently as 20,000 years ago. The range of the woolly rhino (*Coelodonta antiquitatis*) extended from Siberia across what is now the North

Sea into England. Apart from fossil remains, there is also evidence of encounters with early Europeans: a rib bone used for a late Paleolithic carving was found in a limestone gorge in northern England, and the famous rock art in the Chauvet Cave in southern France displays no fewer than 65 depictions of rhinos, presumed to be of the woolly kind. Other caves and Stone Age artifacts also bear witness to the fact that rhinos once were a familiar sight to ancient Europeans.

As the climate changed and human hunters improved their weapons, the woolly rhinoceros disappeared, around 12,000 years ago, along with much of the Pleistocene megafauna, of which only the marine part survives with its species richness almost intact. Like other families of terrestrial megafauna, rhinoceroses vanished everywhere except in parts of Africa and South Asia.

The closest extant relative of the woolly rhino is the Sumatran rhinoceros (*Dicerorhinus sumatrensis*), which is now critically endangered with fewer than 100 animals surviving in the wild. The Javan rhinoceros (*Rhinoceros sondaicus*) and Africa's black rhinoceros (*Diceros bicornis*) share this undesirable IUCN Red List status, while the Indian rhino (greater one-horned rhinoceros; *Rhinoceros unicornis*) is listed as vulnerable and the white rhino (*Ceratotherium simum*) is only near threatened, although its subspecies, the northern white rhino, is extinct in the wild.

Next to habitat loss, the illegal hunting fueled by the demand for the rhino's horn is the biggest concern. Although there is no scientific evidence for any medical benefit and it is chemically identical to clipped fingernails, rhino horn is still valued as a supposed medicine in Asia. Much like elephants and sharks, rhinos are suffering from rapidly growing purchasing power in Asia. On a smaller scale, additional demand for rhino horns comes from Yemen, where ceremonial daggers with handles carved from the horn are prized as a symbol of status and masculinity.

Within the time *Homo sapiens* spread around the world, the family of rhinoceroses moved in the opposite direction, from a globally represented success model to a dwindling group of exotic species. Africa today hosts two of the five surviving rhino species, which are confusingly called black and white rhinoceros, although they are not distinguishable by pigmentation. The "white" one (*Ceratotherium simum*), which is also the largest surviving rhino and thus one of the largest terrestrial mammals, possibly owes its English name to its wide mouth, and the more narrow-mouthed "black" one (*Diceros bicornis*) was simply assigned the opposite attribute to highlight its difference.

Both species only survive in very limited ranges, mostly in national parks. Both have faced extinction risk in the past but are now recovering in numbers. Of the white rhinoceroses, the northern subspecies is extinct in the wild, and attempts to breed offspring from the three individuals surviving in captivity are proving challenging.

The southern white rhino, mainly present in South Africa and neighboring countries, has been brought back from the brink of extinction. There were fewer than 100 of the animals in 1895. Stringent protection within South Africa's national parks has enabled the population to recover and grow exponentially, doubling every decade, such that its size is now estimated at more than 20,000 individuals.

The black rhino is further behind with its recovery. Its most difficult period was between 1970 and 1992, when it was hunted almost to extinction. After a slow recovery, it is still critically endangered with only around 5,000 animals in the wild.

In an attempt to extend the black rhino's range and grow its population, African Parks, based in Johannesburg, South Africa, in collaboration with the Rwanda Development Board, translocated a founder population of 20 eastern black rhinos from South Africa to the Akagera National Park in Rwanda, 10 years after the species was last seen in that country.

African Parks took on the management of the Akagera National Park in 2010 and has since then focused on stamping out poaching and on re-introducing species that have been wiped out in Rwanda. Special provisions have been put in place to protect the new rhino population from any poachers that might try their luck.

Meanwhile, some of Asia's rhinos are also moving to new homes. Nepal's Chitwan National Park, which boasts a growing population of Indian rhinos (*Rhinoceros unicornis*), has repeatedly served as a source to repopulate other nature reserves where rhinos have become sparse, including Bardiya National Park and Shuklaphanta National Park. In recent years, Chitwan National Park has mostly been able to keep poachers at bay. In April 2017, it suffered its first rhino casualty in three years, while security forces were busy in the run-up to local elections.

Elsewhere, the narrow range of the critically endangered Javan rhino (*Rhinoceros sondaicus*) may make it particularly vulnerable to extinction, conservation researchers have warned. The last remaining population of some 60 animals lives in Ujung Kulon National Park, near Mount Krakatoa, Indonesia, an area at considerable risk of earthquakes and tsunamis.

In a study published in 2017, Brian Gerber from Colorado State University in Fort Collins and colleagues used extensive camera traps to monitor the population and found that 62 individuals were alive in 2013, which marks a very slight recovery from a low point of only 25 animals in 1937.[24]

Analyzing the distribution and movements of the population, the researchers found that around 80% of the territory currently used by this last population of the species could become devastated in a 10-meter (33 ft.) tsunami, which in that area is likely to occur within the next 100 years.

The authors therefore suggested that a translocation project should be set up urgently to establish a new population on safer ground. This might also facilitate further population growth, as it is estimated that the habitat in Ujung Kulon National Park has reached its carrying capacity (the maximum population it can sustain). Moreover, the risk inherent in a translocation event will, in this case, be outweighed by the risk inherent in the current situation, the authors argue.

In the presence of the growing demand for the rhino's horn—and the market mechanisms that mean scarcity will only drive prices up and thus profit margins for poachers—scientific methods and creativity are equally needed to stop the slaughter of the rare animals.

There have been various attempts to stain the horns of wild rhinos to make them traceable and less attractive to poachers. In a research paper published in 2018, Cindy Harper from the University of Pretoria, South Africa, and colleagues took the tracing approach to the level of genetic fingerprinting.[25] The authors reported genotype and population analyses of white and black rhinoceroses, enabling forensic scientists to link confiscated horn products to specific poaching incidents and thus to elucidate trade connections and achieve more forceful sentences for those caught.

In a court case in Malawi, three men convicted of poaching a black rhinoceros in Liwonde National Park were sentenced to 18, 10, and 8 years of prison, respectively. A rhino carcass had been discovered in the park on July 13, 2017, and the horns were found the same day using tracking technology. This was the first such sentence since Malawi strengthened its wildlife laws in December 2016.

However, there are limits to what law enforcement can achieve when it faces strong consumer demand—as the failure of the global war on drugs has demonstrated for more than a century now. Some alternative approaches are based on attempting to undermine the market value of the rhino products. Conceivably, flooding the market with synthetic, bioidentical rhino horns could bring the soaring prices back down to the level where poaching is no longer seen as lucrative.

In a detailed analysis of the economics of fake rhino horns, Frederick Chen from Wake Forest University in Winston-Salem, North Carolina, concluded that the most useful approach might be to drive down the quality of the product and thus the price.[26]

"This proposal makes use of a phenomenon in economics known as adverse selection, which occurs when buyers in a market are unable to distinguish between high- and low-quality products. This lack of information can drive down prices enough that high-quality products, which in this case would be real rhino horns, would cease to be supplied by sellers," Chen explained.[27]

As this race to the bottom would naturally be unattractive for businesses producing the fake product, Chen suggests that governments or conservation groups should provide incentives for the production of the rhino-saving substitutes.

Given the precarious state of all but the southern white rhinoceros—which itself has survived a recent population bottleneck—genomic studies are necessary to establish the viability of the critically small populations and to guide any efforts to improve their prospects by breeding and population management.

Mike Bruford from Cardiff University, United Kingdom, and colleagues reported a comprehensive analysis of genetic markers in the black rhino. Using tissue and fecal samples of wild animals as well as skin from museum specimens, the researchers were able to obtain genetic profiles of the surviving populations and those that became extinct within the last 200 years.[28]

The researchers found that the species lost 69% of its mitochondrial genetic diversity in the last two centuries, as many genetic lineages that characterized separate populations in the eighteenth century have become extinct. On the other hand, the authors discovered that the West African subspecies *Diceros bicornis longipes*, which was declared extinct in 2011, survives in a few individuals in the Maasai Mara National Reserve.

The authors believe this detailed mapping of the surviving populations offers the opportunity to rethink conservation efforts. The authors write, "We also identify conservation units that will help maintain evolutionary potential. Our results suggest a complete re-evaluation of current conservation management paradigms for the black rhinoceros." The researchers aim to sequence the full genome of the species next to establish how the genetic bottleneck might affect it. The genome of the white rhinoceros was published in 2012.

In 2018, Herman Mays from Marshall University in Huntington, West Virginia, and colleagues presented the genome of the critically endangered Sumatran rhinoceros (*Dicerorhinus sumatrensis*), enabling insights into the population history of the species as well as its current genetic bottleneck.[29] The results suggest a troubled past even in the Pleistocene, after which anthropogenic pressures pushed the species closer to the edge of extinction. It will be challenging to derive a strategy to bring it back from the brink.

Some rhino saviors never give up, however. Although only three individuals of the northern white rhino survive, Tate Tunstall at the San Diego Zoo Institute for Conservation Research and colleagues reported that the "Frozen Zoo" collection (of frozen cells of rare and endangered, mostly mammalian, species) at the institute holds genetic diversity like that of living rhino populations and would thus be sufficient to bring the species back.

Beyond the somewhat romantic view of rhinos and other big beasts as ambassadors from a bygone geological epoch, there are good ecological reasons to try to save what is left of the Pleistocene megafauna and maybe even bring some of it back. For instance, research has shown that large animals contribute disproportionately to the redistribution of nutrients against the flow of the hydrological cycle and that much of this global nutrient pump has already been destroyed by human activities (see chapter 9).

While the three critically endangered species of rhinos are facing a tough battle for survival and the northern white may be a lost cause, the southern subspecies demonstrates that rhino populations can recover after severe bottleneck situations. Thus, there is still hope for the other four species. Conservation science will have to use all the tricks of its trade to ensure that these rare survivors of the Pleistocene megafauna stay with us in the future.

Extinction in progress

Rhinos, elephants, and many other, less famous, species are now at risk of following the footsteps of those species that have already lost the fight for survival. In 1627 we lost the aurochs (*Bos primigenius*), in 1662 the dodo (*Raphus cucullatus*), in 1936 the Tasmanian tiger (*Thylacinus cynocephalus*)—to name just some of the more widely known species that died out since 1500. The year 1500 is the cutoff for inclusion in the IUCN Red List, which identifies 842 species, subspecies, and varieties as extinct. This number is certain to be an underestimate, however. As of June 2022, the website deadasthedodo.com, for instance, references 5,667 species that disappeared in the last 12,000 years.

In a series of "extinction obituaries," the UK newspaper *The Guardian* is drawing attention to less famous examples of species lost in living memory, such as the Bramble Cay melomys (*Melomys rubicola*), a small rodent from Papua New Guinea declared extinct in 2015 and recorded as the first casualty of anthropogenic climate change; the Hawaiian passerine bird po'ouli or black-faced honey-creeper (*Melamprosops phaeosoma*), which was last seen alive in 2004; and the Christmas Island forest skink (*Emoia nativitatis*), a small lizard lost in 2014.

Hundreds more are currently at risk of following the ones that disappeared. The vaquita porpoise (see chapter 6) and three critically endangered species of rhinoceros (discussed above) are among the most desperate cases treated by conservation scientists, but many others are disappearing without even getting noticed. The EDGE lists prepared by the Zoological Society London are drawing attention to the risks of losing evolutionary unique species, whose disappearance would disproportionately affect biodiversity.

Many experts agree that the ongoing wave of population declines and extinctions is building up to a mass extinction comparable to those we know from the fossil record. A widely used criterion for mass extinctions is the disappearance of 75% of species in a relatively short amount of time by geological standards, typically tens of thousands of years. Although there is some debate about what does and doesn't count, many agree that there have been five mass extinctions so far, making the current biodiversity crisis the sixth extinction, also known as the Holocene extinction.

The idea that we are witnessing and causing the onset of a sixth mass extinction was popularized by Elizabeth Kolbert in the 2014 book *The Sixth Extinction* and has since been widely debated. In 2022, Robert Cowie from the University of Hawai'i at Mānoa, and colleagues at the Muséum national d'Histoire naturelle in Paris, France, reviewed the evidence for the concept of the sixth mass extinction and the arguments of those who aim to deny it or claim it is a natural process.[30]

Those who downplay the current loss of biodiversity or deny that it could be the onset of a mass extinction event tend to focus on the small number of extinctions recorded in the IUCN Red List, which mainly covers vertebrates. As of the 2020 release, the Red List only includes 882 species classified as extinct, which is 0.04% of the 2.2 million recently extant species known to science, or 0.73% of the 120,372 species assessed by IUCN. Based on these figures, some have concluded that extinctions remain within the natural range of background extinction rates to be expected in normal times.

Cowie and colleagues argued, however, that the vertebrates with a Red List assessment are far from representative of all known living species. There are fewer than 70,000 known vertebrate species (3.3% of all species), but, as they are bigger, more visible, and more like us than insects or mollusks, they tend to get more media attention and are more likely to benefit from conservation projects.

Insects, by contrast, represent a much larger share of species diversity, with more than a million species known, but they are not as well represented on the Red List, with only 9,793 (0.9%) assessed for their conservation status. Of those assessed and not found data deficient, only 63 (0.9%) are listed as extinct, but Cowie and colleagues noted that many insect extinctions will go unnoticed and unmourned. As cases in point, the authors referred to several parasite species that are known to have died out along with their vertebrate hosts but were not listed as extinct by the IUCN.

Although insects of bygone times are sometimes preserved as beautiful specimens encased in amber, they are generally less likely to leave identifiable remains than most vertebrates. Moreover, the conservation attention for insects is highly biased in favor of "collectible" species, such as butterflies and dragonflies. Therefore, the authors suggested using a different group of invertebrates as indicators of extinction rates, namely mollusks, of which there are more than 83,000. Many mollusk species can be classified by their shells even after their demise.

As of June 2022, the IUCN has evaluated just over 10% of known mollusk species and records 299 as extinct, which is 4.6% of those that aren't data deficient. Land snails from Pacific islands account for a large proportion of these extinctions.

Cowie and colleagues compiled further mollusk extinctions reported in the literature and arrived at 638 species extinct, 380 possibly extinct, and 14 extinct in the wild—a total of 1,032 species. They argued that the IUCN classification tends to underestimate extinctions for fear of committing the "Romeo error" and having to backtrack when a species is discovered to still be alive, and therefore Cowie and colleagues go with their own higher figure.

Boldly extrapolating these findings to all of biodiversity, Cowie and colleagues estimated that between 7.5% and 13% of all species disappeared within the last 500 years. This result, the authors suggested, is more likely to be accurate than any rates determined for vertebrates or insects. It is several orders of magnitude higher than the accepted background rate and the proportion of extinct entries on the Red List, which is 0.04%.

If the rate persists on the timescale of millennia, as it may well do because of the permanent changes caused by human intervention, it is entirely plausible to predict that the threshold of 75% species loss will be crossed in a relatively short time compared to previous mass extinctions. Cowie and colleagues suggested defining the onset of the extinction event as the point when human hunters emigrated from Africa and started to cause extinctions among the megafauna of other continents, between 200,000 and 45,000 years ago.

Further detailed analyses suggest that the marine realm has so far been spared this wave of excess extinctions. Although the size and inaccessibility of the marine environment adds an element of uncertainty to any estimates, it is plausible that these factors, in limiting human-inflicted damage, have also saved marine biodiversity. As industrialized fishing and exploitation of seafloor minerals only got underway in the twentieth century, by which time awareness of environmental problems has also become prominent driver of events, there is hope that much of the marine biodiversity can be saved, even though ocean acidification and climate warming are posing specific threats to marine ecosystems, including coral reefs. We will come back to these marine issues in the next chapter. A less severe impact on marine diversity would set the ongoing mass extinction apart from the previous five, which affected marine species as much or more than terrestrial ones.

Plants are also in a better position than terrestrial animals. Although many are threatened by habitat loss, they are much easier to save (e.g., by seed storage and breeding in botanic gardens) than animal species in a similar situation.

Islands have been most severely affected by extinctions after the arrival of human hunters and settlers. This is especially true for remote islands, such as the Hawaiian Islands, and many archipelagos in the South Pacific, which have had millions of years to develop their own characteristic diversity of species, many of which (like the Hawaiian po'ouli mentioned above) have fallen victim to the arrival of humans, who were accompanied by rodents, cats, and other invasive species. Across all Pacific islands, more than 2,000 bird species are estimated to have died out since humans arrived, representing a significant part of the global bird diversity.

Conservation is still very much focused on large and charismatic vertebrate species, such as elephants and orangutans. Where unique human threats such as hunting are the main danger, addressing this problem can ensure the survival of a species. Some of those species can also be bred in captivity and rewilded when conditions have improved, as has been successfully demonstrated for the Arabian oryx and, with a little help from cloning technology, for the black-footed ferret.

These laudable efforts don't scale to the threat of a mass extinction, however, which may eventually wipe out 75% of all existing species. As Cowie and colleagues warned, any success stories should not mislead us to believe that we are successfully combating the biodiversity crisis.

For the sake of efficiency in tackling a near impossible task, conservation work must improve conditions for many species—for example, by stopping habitat degradation in environments known as biodiversity hotspots and by stopping climate change. "Despite the rhetoric about the gravity of the crisis, and although remedial solutions exist and are brought to the attention of decision-makers, it is clear that political will is lacking," Cowie said. "Denying the crisis, accepting it without reacting, or even encouraging it constitutes an abrogation of humanity's common responsibility and paves the way for Earth to continue on its sad trajectory towards a Sixth Mass Extinction."[31]

Although the debate over numbers and definitions is likely to continue, it is abundantly clear that habitat protection must be stepped up considerably and on a global scale. A timely reminder of the scale of efforts needed came from James R. Allan of the University of Amsterdam in the Netherlands and colleagues, who calculated the amount of land surface that needs to be protected if the biodiversity crisis is to be stopped.[32] The researchers found that 64 million square kilometers (24.7 million mi.²), or 44% of the continental surfaces, would require some sort of conservation action, with measures ranging from protected areas to land-use policies. More than 1.8 billion people live within the areas in need of some protection, so reconciling the interests of nature conservation with those of the human residents is an important challenge if such dramatic action is to go ahead.

Can we change our predatory ways?

On July 1, 2015, the Wildlife Conservation Research Unit (Wild-CRU) at the University of Oxford, United Kingdom, lost one of its study subjects. Researchers from the unit have studied and tracked lions in Hwange National Park in Zimbabwe since 1999, and have tagged 62 lions in that area, of which 34 have died, including 24 at the hands of sport hunters. As Zimbabwe allows around 40 to 50 licensed lion killings per year, it's not that unusual for a lion to be killed in Zimbabwe.

That July, however, the killing of one adult male lion tagged by WildCRU stirred world media attention for weeks. As reports emerged that 13-year-old Cecil, a long-standing favorite with visitors to the park, had been coaxed out of the protected area and then wounded and tagged with an arrow and finally shot dead with a rifle, there was a global outcry (certainly helped by the seasonal shortage of political news).

The ill-advised hunter, a US dentist, managed to return home but had to go into hiding and keep his practice closed as the anger of animal lovers erupted. One of his local guides now stands trial for illegal hunting and could face up to 15 years in prison. Meanwhile, WildCRU may find consolation in the hundreds of thousands of dollars in donations the center received from sympathetic animal lovers mourning the death of Cecil.

In the bigger ecological picture, Cecil's case is just one of many examples showing that we humans are far from enacting our role as top predator responsibly or sustainably. Like that trophy-hunting dentist, we are killing the wrong animals all the time—many of them for food, some just for kicks.

Hominins joined the ranks of top predators very late and somewhat unexpectedly. We never evolved sharp claws or long canines to tear prey apart. Unlike the ancestors of our domestic cats and dogs, our own primate forebears mostly ate plant food until around two million years ago, like chimpanzees still do today. Only when *Homo erectus* conquered the African savanna and evolved to become a long-distance runner and competent spear thrower did hominins become a significant predator.

The first hunter-gatherers gradually improved their weaponry and came to depend on meat as energy-rich fuel for their expanding brains. The use of fire for cooking, although not proven conclusively for that time, may have helped. Considering that the savanna is a biotope shaped by fire and burns regularly, it is plausible to assume that *Homo erectus* learned to use fire quickly, even though the natural fires will have erased the evidence.

Animal species that coexisted with the emerging *Homo sapiens* hunter-gatherers in Africa had time to adapt to the new threat—which is the likely reason megafauna there survived for longer than elsewhere. When hunters spread to other continents, they caused significant extinctions of the local megafauna.

Even today human hunters are still a significant threat to some of the surviving animal species. Part of the problem is that our hunting and fishing activities are highly unusual and don't fit the ecological role of a normal apex predator, as Chris Darimont from the University of Victoria, Canada, and colleagues showed by comparing the casualties of human hunting and fishing to the prey selection of other predators.[33]

The global data analysis shows that humans exploit adult prey at a much higher rate than other predators. While this preference is understandable in terms of hunting or fishing efficiency, it is in many cases unsustainable as it reduces the reproductive capital of the prey population. Other predators, by contrast, would be more likely to target juveniles. In a somewhat heartless ecological analysis inspired by economics, this feeding strategy is comparable to taxing the interest that the capital produces while leaving the capital intact.

The unnatural preference for large prey is driven to its perverse extreme in the case of trophy hunters who explicitly seek out the largest individuals in order to have more impressive specimens to display on their walls. As animals living in groups, like lions or wolves, depend on the largest individual for the social order in their group—side effects of the killing of that one impressive "trophy" animal may result in further deaths due to group instability. Moreover, the increased death risk of larger specimens may in some species act as a significant selection pressure and thus alter the evolutionary trajectory of the species.

While the hunt for terrestrial animals has become a marginal activity compared with agricultural meat production, which has its own set of side effects to be discussed below, the industrialized fisheries industry shows the same anomaly, removing a much larger portion of adult individuals in a prey population than any other predator would.

Darimont and colleagues concluded that "humans function as an unsustainable 'super predator,' which—unless additionally constrained by managers—will continue to alter ecological and evolutionary processes globally." The authors suggested that the typical kill rates of predator species should be used as models to establish sustainable limits for human hunting and fishing, which would in many cases be much lower than the rate of killing that our species inflicts today.

After nearly two million years of hunting and gathering, humans evolved a new way of exploiting plants and animals: agriculture. Paradoxically, the benefits of this development for the individual are far from obvious, and it is likely that the ability of agricultural methods to feed more people per area and sustain a more stratified society, including monarchs and soldiers, explains why agriculture spread around the world from just a few nucleation areas.

Although it may look like a more peaceful and sustainable alternative to hunting, farming animals brought its own set of problems. The first drawback that early farmers had to cope with was the emergence of zoonoses—diseases that jumped from animal hosts to the humans who lived in proximity with them (see chapter 11). Some of these pathogens have co-evolved with their human hosts to become less deadly and now form the widespread "childhood diseases," which are rarely fatal in populations that have lived with them for many generations but may endanger the last remaining hunter-gatherer tribes that lack immunity to them. Others, like avian influenza strains emerging in Asia, are still causing concern today.

The rapid growth of industrial-scale meat farming has caused additional problems. Average meat consumption per capita has almost doubled in the 50 years between the mid-1960s and the 2010s. Multiplied by population growth, this gives a fivefold increase in global meat eating. On average, people in the United States (121 kg, or 267 lb., per person per year) and Australia (114 kg, or 251 lb.) are the most ardent meat consumers.[34] The UN Food and Agriculture Organization (FAO) figures for 2017 are based on meat available to eat, not accounting for what goes to waste. The perceived association of a meat-rich diet with wealth is certainly part of the psychological problem, with the richer nations enjoying it as a reward experience and the poorer ones aspiring to share more of that reward.

According to FAO figures, gases released by farm animals are already responsible for 14.5% of the anthropogenic greenhouse effect and thus a major driver of climate change. The expansion of meat production also drives deforestation, adding again to climate change, as well as the unsustainable exploitation of water resources.

Water experts have warned that the world is on track for a severe water shortage. A report released by the Stockholm International Water Institute in 2012 predicted that, if we continue in business-as-usual mode, water resources will not suffice to feed the anticipated world population of nine billion people in 2050.

The rapid growth and relentless drive to higher efficiency of meat production has led to industrial production methods that are far removed from anything that could be considered a natural lifestyle for an animal. Animal welfare and issues of environmental and food security have long served as key arguments in the quest to curb humanity's hunger for meat, but are we ready to change the menu?

If the first hunter-gatherers spiced up their plant diet with meat and thus gained extra energy to fuel their expanding brains, it's not completely unreasonable to expect that modern humans might use that fabulous brain to find ways out of their meat addiction.

Eating as much meat as Americans and Australians do is clearly unhealthy and has been associated with the rising incidence of bowel cancer and type 2 diabetes, as well as other diseases prevalent in the wealthier nations. Health organizations frequently remind us that we all should eat more fruit and vegetables and less meat, as do environmental organizations, without making a significant dent in the statistics so far.

Alternatives for those who want to keep eating meat-like textures and flavors exist in the shape of soy and fungal protein products. Some producers are trying to break out of the vegetarian and vegan niche market by presenting their food as a healthy, fat-free alternative.

From the environmental angle, much could be gained if more people chose to eat insects instead of vertebrates. But still, the widespread psychological associations of meat with wealth and plant food and insects with poverty are making it all but impossible to make these options attractive to those who can afford to eat meat.

To overcome these deeply ingrained perceptions, campaigners will have to make use of the same hedonistic instincts that currently drive people to indulge in eating meat. As Ophelia Deroy from the Centre for the Study of the Senses at the University of London, United Kingdom, argued in an opinion piece, it is important that green alternatives like insect-based food are appraised with their genuinely appealing sensory qualities, such as their crunchy textures or spicy flavors.[35]

Modern-day consumers won't change their ways in significant numbers, Deroy argued, if they're told they have to in order to save the planet. They might, however, if a new food option appears more appealing than what they used to eat. Thus, fried house crickets should not become the next wartime rationing, but they should be seen as the next culinary craze following in the wake of the sushi outlets.

Only very occasionally restaurants take up that challenge. In 2021, however, Australia's national science agency, the Commonwealth Scientific and Industrial Research Organisation (CSIRO), launched a roadmap for the nascent insect-food industry, which provided "a comprehensive plan for the emerging industry, exploring the challenges, opportunities, cultural values, sustainability, and health outcomes of the edible insect industry in Australia."[36] Many more initiatives like this one will be necessary to finally stop our omnivorous species from hunting, fishing, farming, and eating the wrong kinds of animals.

CHAPTER 6

Save Our Seas

After the doom in the previous chapter, here is some good news: the oceans aren't as badly affected by our misdeeds as the continents—yet. A total wipeout of several species of whales was averted at the very last minute, when most of the world banned industrial-scale whaling.

This has left us with the curious situation that the oceans have kept their Pleistocene megafauna, whereas continents (except for Africa), for the reasons discussed in the previous chapter, have lost it almost completely. This discrepancy is easily explained by the difficulty of hunting big game in the oceans. Until commercial whalers came along, the marine megafauna was protected by the sheer size of the oceans and poor accessibility of the open-sea habitat.

For the same reason, however, we know far too little about marine ecosystems, and virtually nothing about the deep seafloor. What we do know, however, is that the excess carbon dioxide that causes climate change also enters the seawater causing ocean acidification and serious threats to multiple species. Moreover, rising sea levels can drown coral reefs and important coastal habitats.

Therefore, we need to understand the oceans much better to ensure that our various forms pollution of (including sound pollution) don't end up killing off the marine megafauna as well as sensitive marine ecosystems such as coral reefs.

Can science rescue coral reefs?

It's not easy being a coral in the ocean. The complex lifestyle of the polyps and their algal symbionts depends on a range of parameters being just right, including sunlight, temperature, and pH. Even though the polyps are animals, they are immobile in their adult life, so they can't flee if environmental conditions become dangerous.

At the end of the last ice age, sea levels rose dramatically, leaving many coral reefs to "drown" at a depth where they no longer received sufficient sunlight to grow. This drastic change favored the fast-growing genus *Acropora*, which had been insignificant until then but became the dominating species around the tropics thanks to its ability to grow as fast as the sea levels rose, as research from Wolfgang Kiessling's group at the University of Erlangen-Nuremberg, Germany, has shown.[1]

While this ability is still an asset today, as sea levels are again rising, there are other aspects in which *Acropora* is not as well equipped to deal with environmental change. It is sensitive to the prevalent threats that are all linked to CO_2 emissions from the burning of fossil fuels; namely, ocean acidification, chronic warming of the water, and acute warming in El Niño events (characterized by anomalously high temperatures in the tropical surface waters of the Pacific), which more often become extremely strong.

When the water stays too warm for too long, the co-operation between polyp and symbiont breaks down, as the polyp expels the photosynthetic organism—an event that is known as coral bleaching.

One of several possible triggers for heat-induced bleaching is the increased oxygen production of the symbiont. Catastrophic widespread bleaching has been observed more and more frequently since the late twentieth century. One of the most devastating episodes occurred in 2016 and affected Australia's Great Barrier Reef among others. Terry Hughes from James Cook University, Townsville Campus, Australia, and colleagues analyzed the distribution of bleaching effects and found it pointed to higher water temperatures with climate change as the primary cause.[2]

Bleaching is often linked to a strong El Niño event, and the 2016 event was linked to the strongest recorded El Niño event so far. Although there is no proven mechanistic link to climate change, strong El Niño events have become more frequent in recent years and may end up becoming a regular occurrence.

After previous bleaching episodes, reefs that were heavily affected took around a decade to recover. If bleaching on a global scale were to happen several times in a decade, this might spell the end for many tropical coral reefs, which in turn would have apocalyptic consequences for marine biodiversity and for fisheries. Experts have predicted that we may lose the reefs by the middle of this century—unless science can come up with a revolutionary way to save them.

The very first thing required in any attempt to save the coral reefs is information about the diversity, sensitivity, and resilience of the existing coral populations. Given the diversity of corals, their symbionts and their ecological context, the handling of the required information alone is one of biology's "big data" challenges.

Although the major bleaching events have been studied in some detail, it is still far from clear which factors determine whether a given population will survive a certain exposure to environmental stress. And only when researchers understand why some corals survive better than others will they be able to move onward to trying to protect all corals.

To allow a better use of the existing information on coral bleaching, Luisa Marcelino's group at Northwestern University in Evanston, Illinois, systematically analyzed and compiled published information regarding nearly half the world's corals, covering 316 sites, to create a "Global Bleaching Response Index." The researchers developed a quantitative measure of the temperature sensitivity, assigning a numerical value to each coral species.[3]

Tracy Ainsworth from James Cook University in Cairns, Australia, and colleagues studied thermal resistance of corals on the Great Barrier Reef by analyzing records of past bleaching and die-offs along with 27 years of satellite data on surface temperatures. They found evidence suggesting that, in the past, mechanisms of resistance induced by less pronounced, sublethal warming before the main event have helped the corals survive. However, they conclude from their analyses that these mechanisms will fail to function as soon as average water temperature has risen by an extra 0.5°C (0.9°F), meaning that the Great Barrier Reef will soon become more sensitive to El Niño events than it has been so far.[4]

One small hope is that at least some coral species may be able to evade their doom by moving—on the timescale of generations. The Great Barrier Reef, which spans several climate zones, offers examples of different species adapted to different temperature regimes. On the timescale of decades, the species distribution may move along with shifting climate zones. The concern is, however, that human-made climate change happens on a much faster timescale than previous natural changes, and it may be too fast for this kind of adaptation to work. At the very least, it will not work for the entire ecosystem as it existed before climate change—like the sea-level rise at the end of the last ice age, it will produce selection pressure changing the composition of the coral communities.

Another potential way out is the poorly understood world of coral reefs in deeper waters, the so-called mesophotic (twilight) coral ecosystems (MCEs). A 2016 report commissioned by the United Nations and launched at the UN Environment Assembly in Nairobi, Kenya, investigated whether species losing their habitat in surface reefs could find refuge in the more stable and shielded environments offered in the twilight zone of the oceans.[5]

The report noted that MCEs "remain largely unexplored in most parts of the world, and there is little awareness of their importance among policy makers and resource managers." With this report, the researchers and the United Nations Environment Programme (UNEP) aim to raise this awareness in the context of the imminent threat to the survival of more widely known shallow-water reefs.

The answer to the question whether MCEs can help to rescue species from declining surface reefs depends on the species concerned and the environmental situation. The report concluded: "In some situations, MCEs may provide this ecosystem service and act as 'lifeboats' for nearby, connected shallower reefs that have been damaged. In other cases, however, MCEs may be just as vulnerable as shallower reefs to the range of human pressures exerted upon them."

While MCEs are better protected from some threats, they are still vulnerable to others, such as being covered by sediment runoff or disturbed by resource extraction from the seafloor, trawling, or cable laying. Lack of protection is often due to a lack of awareness of the very existence of these hidden reefs, which is why the report's main aim is raising awareness.

What, if anything, can conservation workers do to rescue corals? One could consider helping the corals move to higher latitudes, as subtropical waters will become more suitable for the species currently thriving in the tropics, as Erica Hendy and colleagues at the University of Bristol, United Kingdom, have shown.[6] Conservation workers could support relocation, for instance, by providing artificial reef substrates to make new habitats accessible and bridge gaps which corals might not naturally overcome, or even by moving them manually to speed up a range shift that might otherwise be too slow to guarantee survival.

A more drastic intervention is championed by coral researchers at the Hawai'i Institute of Marine Biology. The institute's director, Ruth Gates (1962–2018), and her group were building on the natural selection enacted by two major bleaching episodes, which hit their local reefs in subsequent years, in 2014 and 2015. Gates and her team studied the genetic and epigenetic traits of the surviving corals with a view to breed new strains that might be better equipped to withstand the coming coral crisis. Media reports have sensationalized Gates's quest as the breeding of "super corals"; Gates embraced this term and went on to use it herself. Since her untimely death, work at the Coral Resilience Lab has continued to build on her legacy.

Modern genetic analyses and gene manipulations make the breeding of a super coral the most straightforward part of the venture. Initial experiments using temperature changes to "train" corals to trigger epigenetic adaptation, and others attempting to make their algal symbiont more stress resistant, showed promise early on. Together with Madeleine van Oppen from the Australian Institute of Marine Science in Townsville, Gates established that corals using clade D algal symbionts would be more resistant to warming, but the corals investigated preferred the clade C algae, which produce more food. By selective breeding, the researchers attempted to make the preferred symbiont more stress resistant.[7]

In addition to having the internalized algal symbionts, corals co-operate with an external microbiome populated by a little-characterized diversity of bacteria. Gates and van Oppen aimed to analyze the microbiome of all existing coral ecotypes to find additional mechanisms that may help to determine the resilience of corals under environmental stress.

Once resistant super-coral strains supported by optimized algae and bacteria exist, the trickier part will be to seed them on a scale commensurate with the reefs that are being lost to climate change. Currently, researchers can establish artificial coral settlements on the scale of meters—but if they want to replace what the tropical oceans are set to lose, they need to develop technologies that work on the scale of thousands of square kilometers.

And then there are some deep philosophical issues around the question of what exactly conservation is about. For those concerned only with the ecosystem services and particularly with the benefits for the global economy, lab-bred replacement corals will be fine as long as they are functional.

For everybody who wants to conserve at least some of the last traces of the natural world as it existed before humans conquered the globe, the introduction of super corals may appear as the marine equivalent of replacing buffalo with cattle or wolves with dogs. It may look similar and work the same way, but it will still be artificial.

Among the many species threatened by the predicted decline of coral reefs are the 28 different species of clownfish (subfamily Amphiprioninae), each adapted to life within a certain species of sea anemone. Notwithstanding their floral name, sea anemones are animals related to coral but lacking the mineral structures. They, too, can suffer bleaching at higher temperatures.

The symbiotic relationship between sea anemones and their resident fishes has been studied intensively—most recently in a cooperation between Vincent Laudet and Noriyuki Satoh at the Okinawa Institute of Science and Technology in Japan, who studied the symbiosis from the perspective of the anemone's gene expression.[8]

As these fish have been universally popular since the release of the Disney-Pixar movie *Finding Nemo* and its sequel, awareness of their dependance on anemones could well be a wake-up call that the world will listen to. If we want to save Nemo, we will have to save the sea anemones and the coral reefs.

Life's islands under the sea

In September 1835, the HMS *Beagle* reached the Galápagos Islands, where Darwin spent five weeks exploring the local wildlife. He found it quite different from the nearest continental mainland, Ecuador, which is around 1,000 kilometers (620 mi.) away. Mammalian predators and insects were conspicuously absent. On the uninhabited islands of the archipelago, wildlife was unafraid of human visitors and thus easy to study, catch, or shoot. As was usual procedure at the time, the *Beagle* crew picked up giant tortoises to stack them in the hull of the vessel as live meat provision.

Birds were thriving in a confusing variety of closely related species, including the finches that Darwin was to study in detail based on the specimens he shot during his visit. These observations were to become a key foundation of his nascent theory of evolution of species, where he emphasizes the role of geographic barriers for the divergence of species.

After Darwin's realization that species change over time, it took another century for humanity to become aware that our planet also gradually changes as continents move around, disrupting connections and creating new ones. Sea-level changes caused by climate change have also redrawn the map as recently as the end of the last glacial period, when Britain became an island. The Galápagos archipelago, appearing out of the ocean due to volcanic activity just a few million years ago, may be an extreme example, but even on the continental land masses, geography changes all the time, typically on a time scale like that of the evolution of species.

Establishing the connection between geographic and biological change, the pioneers of island biogeography set out to establish rules to understand how the isolation of islands influences the fate of species. Well ahead of his time, Georges-Louis Leclerc, Comte de Buffon (1707–1788), had speculated that the "Old World" and "New World" had once been connected. Darwin and his contemporaries then went on to link evolution to geography, although they were still hindered by their lack of awareness of continental drift. Championed by Alfred Wegener (1880–1930) since 1912, the idea that continents move around only became widely accepted after the 1950s, when plate tectonics provided a plausible mechanism for their movements and subsequent seafloor exploration revealed the spread and subduction zones where the ocean floor grows and shrinks, respectively.

With these foundations in place, Robert MacArthur (1930–1972) and E. O. Wilson (1929–2021) proposed the term "island biogeography" in 1963 and developed what is still regarded as its standard model. They had conducted experimental fieldwork, which included clearing small islands in the Florida Keys of all insects and then observing their resettlement.[9] Based on their observations, MacArthur and Wilson identified colonization and extinction as the main determinants of species richness on an island. Geographic parameters like distance from other landmasses and size of the island can impact these drivers. For instance, an island larger in diameter offers a wider target for organisms randomly dispersing from another landmass, so it facilitates colonization. An island larger in surface is likely to offer better survival chances for a larger number of species, thus reducing the likelihood of extinction.

With these fundamental concepts, ecologists have gone on to study biodiversity in many parts of the world, not just on islands. The conceptual framework has proven useful for isolated habitats on land, such as mountain tops or lakes, as well as fragments of nature cut off by artificial structures like roads or agricultural land. Even underwater, marine biologists can find islands of life following variations of these rules.

Water around islands is not the only medium that can isolate a species. On the continents, natural environments like oases in the desert, highland biotopes surrounded by lowlands, caves, and lakes under the ice shield of Antarctica can act as islands for a segregated set of species. As the barriers between such areas may be surmountable for some species but not for others, the degree of isolation tends to depend on the kind of species studied.

Conservation science soon realized that island biogeography also applies to issues like forest fragmentation, where the surrounding matrix, impenetrable for many forest specialists, can be agricultural land or savanna. The Biological Dynamics of Forest Fragments Project (BDFFP), a long-running ecological experiment in Brazil, where forest land was fragmented and studied over decades, has shown that very similar considerations apply in this situation as on islands (see chapter 4).

Other ecological islands on land include areas cut off by the dramatically accelerating global infrastructure development that will see 25 million kilometers (15.5 million mi.) of road built between 2010 and 2050, a 60% expansion of the existing network. Similarly, large hydroelectric dam projects interrupting the continuity of river systems can convert them into chains of isolated lakes by making migration difficult or impossible.

The results from island biogeography studies and fragmentation analyses like BDFFP have revealed that for efficient conservation, larger, undisturbed reserves are better. This philosophy has also informed the installation of wildlife corridors connecting protected areas, like the jaguar corridor established in 2012 to connect the big cat's territories in Central and South America. On a local scale, small measures like wildlife bridges or tunnels across roads and fish ladders bypassing dams can help to avoid further fragmentation of wildlife habitat.

How protected areas can best guard species is also an urgent question in the marine realm. Does island biogeography also apply underwater, and can it help marine conservation?

The marine environment includes many potential ecological islands, such as coral reefs, shallow-water biotopes around islands and seamounts, and the surprisingly rich biotopes surrounding warm springs and hydrothermal vents at the ocean floor. While all of these are isolated due to being surrounded by a matrix with clearly different conditions, the fluidity of the matrix and the lack of clear boundaries complicate the situation, meaning research on submarine ecological islands is relatively sparse.

Seafloor communities thriving in the extreme heat and unusual chemical conditions around hot springs and hydrothermal vents provide a dramatic example of isolation in aquatic environments. Their habitats only have a limited life due to the variability of geological activity, and they tend to be hundreds of kilometers away from similar sites. And yet, similar communities have been described in distant locations, leading researchers to conclude that the random distribution of large numbers of larvae with the ocean currents must help to maintain exchange over long distances.

Conversely, Shana Goffredi and colleagues from the Monterey Bay Aquarium Research Institute in Moss Landing, California, described two hydrothermal vent fields located just 75 kilometers (47 mi.) apart within the Gulf of California but hosting very different fauna.[10] As it turns out, the different geochemistry of the vents favors very different groups of species. The Pescadero Basin vent field, discovered in 2015, is different from most known sites in that its fluids pass seafloor sediment and convert biomaterial into methane and other hydrocarbons. The fauna described at this site consists mainly of worm species, many of which were new to science.

The authors noted that these findings have implications for conservation concerns associated with the nascent industry of deep-sea mining. As of 2022, after a few delays and false starts, it appears likely that commercial operators will soon start using heavy robotic equipment to harvest minerals containing metals such as copper, zinc, silver, and gold from deposits accumulated around such sites. The operations risk destroying the local fauna, so the decision whether they are permissible depends on the prospects of recolonization of the sites after the minerals have been harvested.

Closer to the sea surface, on the background of the current crisis threatening the loss of most coral reefs due to ocean warming, coral researchers have studied the dispersal mechanisms of corals more closely. As we have discussed in the previous section, migration to higher latitudes may save some coral populations from extinction, but a lack of suitable submarine substrates means that humans may have to help them break out of their insular confinement.

Like the corals themselves, reef fishes may also be confined by the insular habitats provided by reefs. However, ecological islands under the waterline have so far been neglected by island biogeography, according to Hudson Pinheiro from the California Academy of Sciences in San Francisco and colleagues. These researchers studied the evolution of 10 endemic fish species at the Vitória-Trindade Chain in the South Atlantic, off the coast of Brazil.[11] This is a volcanic ridge that, at current sea levels, features shallow reefs and two small islands only at its eastern end (more than 1,200 km, or 745 mi., from the mainland), while several former islands are present as seamounts between 20 and 60 meters (66 and 200 ft.) below the sea surface. At lower sea levels in the Pleistocene, there were up to 11 islands forming a chain toward the mainland, with shallow-water habitats only separated by 100 kilometers (62 mi.) of deep water.

Due to their extremely isolated location, low biodiversity, and well-characterized geological history, these islands are a useful natural laboratory to test the predictions of island biogeography for underwater life. Analyzing the haplotypes of 10 endemic species from the island complex and from seamounts in the chain, Pinheiro and colleagues could reconstruct how immigration from the mainland at low sea levels and extinction shaped the biodiversity of these isolated locations. They found that 7 of the 10 endemic species studied have colonized the islands during the Pleistocene, presumably taking advantage of the sea-level fluctuations and the emergence of island stepping stones between the current islands and the mainland.

Only three species, referred to as paleoendemics, arrived in earlier epochs (Pliocene or Miocene) and survived Pleistocene climate change, while their relatives on the coast of the mainland appear to have gone extinct. While island locations generally carry a high risk of extinction, this finding shows that they can also serve as a refuge for some species.

Analyzing their findings in the framework of island biogeography, the authors note that extreme isolation effects seen in terrestrial island biotopes are less likely, as dispersion is easier in the fluid medium, and most niches are rapidly filled by efficient dispersers. Only thanks to the unique emergence of stepping-stone islands at low sea levels in the Pleistocene could weak dispersers colonize the shallow waters of the island chain and become endemic there. While the general rules of island geography still apply underwater, the authors concluded, their interaction with geological and climate change affect marine species in different ways compared to terrestrial ones.

A much more complex situation is found in the "Coral Triangle," the area between the Indian and Pacific Oceans, which hosts the highest marine biodiversity known. Biogeographers have long debated whether this wealth of species is due to particular productivity in speciation or to an overlap of species ranges originating from both oceans. In other words, depending on the direction of species dispersal, the area could be a source or a sink of biodiversity.

In a study using population-level DNA analyses to map the evolution of the species range of coral reef assemblies, Martin Genner's group at the University of Bristol, United Kingdom, found that the longer-established populations tended to be closer to the heart of the Coral Triangle, suggesting that it acts as a source rather than a sink.[12]

A better understanding of marine biogeography is crucial if we want to save the uniquely rich marine fauna, which has so far escaped the Holocene mass extinction inflicted on terrestrial species by the rapid expansion of *Homo sapiens* (see chapter 5). The beginning of industrialization of resource extraction from the oceans and the seafloor is threatening marine biodiversity and requires protection based on a profound understanding of how marine communities are connected to each other and to their environment.

In September 2016, the World Conservation Congress, held in Hawaii by the International Union for Conservation of Nature (IUCN), passed a resolution calling on governments to set apart 30% of oceans as "highly protected" areas. until then, only around 2% of the oceans were protected. However, China and Japan were among the minority of countries opposing this resolution as too ambitious. On the other hand, E. O. Wilson argued that 30% is not enough and called for half the ocean area to be protected.

Save our seagrasses

Around 50 million years ago, an animal species vaguely looking like today's hippopotamus took the plunge and became fully aquatic. Its descendants are now widely known as whales and dolphins. At around the same time, several lineages of flowering plants (angiosperms) were moving in the same direction, back into the oceans. They became the seagrasses, a grouping of around 60 fully marine species that can pollinate underwater.

The demands of plant life in coastal waters, requiring photosynthesis, pollination, root systems, and a canopy resistant to the shear forces of waves and currents, have led to several traits evolving independently in the separate plant lineages that returned to the oceans. For instance, the blade-like shape of the leaves that led to the name seagrass is an adaptation to water forces and does not indicate relatedness—a typical example of convergent evolution.

Therefore, the diversity of seagrasses has confused botanists since the beginning of systematics. With the help of some molecular studies, the 60 known species can now be grouped into four families (Posidoniaceae, Zosteraceae, Hydrocharitaceae, and Cymodoceaceae), some of which also include non-marine relatives. All four families belong to the order of Alismatales.

Growing in the shallow waters along the coasts, seagrasses are often affected by human activities, including water pollution and development of coastal structures such as harbors. It was only recently, partly due to the growing awareness of the ongoing climate catastrophe, that their ecosystem services came to be appreciated. Among providing other important services, these inconspicuous marine meadows sequester and store significant amounts of carbon.

When it comes to redressing the global carbon balance, every green plant contributes by pulling carbon dioxide out of the atmosphere (or the ocean) and converting it into biomass. The question is what happens to the biomass afterward, and in this respect, it looks like the marine meadows may be particularly useful—and their loss especially dangerous.

Cristian Salinas from Edith Cowan University, Joondalup campus, Australia, and colleagues analyzed the loss of seagrass meadows in Cockburn Sound, Western Australia, and its impact on carbon stock in the soils.[13] This enclosed coastal basin of around 124 square kilometers (48 mi.^{2}) is located 25 kilometers (15 mi.) south of Perth and hosts 10 species of seagrass, with the genus *Posidonia* dominating. Industrialization of the area since the middle of the twentieth century has led to eutrophication (nutrient enrichment leading to excessive algal growth and oxygen depletion) and the loss of 23 square kilometers (9 mi.^{2}) of seagrass meadow between mappings made in 1967 and 1999, reducing it to one-fifth of its original extent.

Salinas and colleagues took soil cores from 11 areas still covered by *Posidonia* seagrasses and from 13 that were bare but known to have been vegetated in the past. Analyzing the carbon stocks in the soil, the authors found that vegetated soil in shallow water, at depths of less than five meters (16 ft.), stored 4.5 kilograms (10 lb.) of organic carbon per square meter (approximately 11 ft.^{2}), while bare soil only had a tenth of the carbon content. In deeper water, however, they detected no significant difference.

The authors concluded from this finding that seagrasses in shallow water strongly protect the soil from being disturbed by waves and currents. When the seagrass coverage is removed, the carbon stored in the soil is lost within a decade, partly due to erosion. For climate implications, this means that the carbon once stored in seagrass meadows that died more than 10 years ago is already gone. Restoring these meadows in the same location thus has no advantage over planting seagrass elsewhere.

Extrapolating the findings to the whole of the Australian coast, the authors calculated that the loss of seagrass meadows constitutes a significant part of Australia's carbon emissions from land-use change, contributing around 2% of the total. Thus, stopping the ongoing decline of seagrass meadows is the most effective action to support. However, planting new seagrass also has a significant carbon benefit, and its carbon accumulation rate in soil compares favorably with that of many land plants. "Seagrass meadows have been estimated to store CO_2 in their soils about 30 times faster than most terrestrial forests," Salinas said. "Seagrass meadows have been under constant threat in Australia through coastal development and nutrient run off since the 1960s. On top of that climate change is causing marine heat waves that are catastrophic to the seagrasses."[14]

On top of carbon sequestration, the seagrass offers many other benefits for the coastal environment, not least protection from erosion. Data from the Cockburn Sound study showed that seagrass helps stabilize sediment in shallow water. Apart from retaining carbon stock, this influence is also valuable in fighting the loss of beaches to coastal erosion. Beaches, especially in the tropics, are important both for wildlife such as sea turtles and for the tourist economy. Many tropical beaches are at acute risk of erosion. Where funds are available, engineering solutions such as seawalls and replenishment of sand have been tried, but they are likely to cause ripple effects elsewhere.

Rebecca James at the Royal Netherlands Institute for Sea Research in Yerseke and colleagues studied the effect of seagrass on coastal sediments at Galion Bay, on the Caribbean island of Saint Martin, experimentally using a portable and adjustable field flume to manipulate water flow and observe the effects on the sediment.[15] "We showed that seagrass beds were extremely effective at holding sediment in place," James said. "Especially in combination with calcifying algae that 'create their own sand,' a foreshore with healthy seagrass appeared a sustainable way of combating erosion."[16]

For tropical countries that are unable to pay for expensive protection structures and are losing beaches to erosion as a result, this combination appears to offer an affordable solution. It helps to protect tourism income as well as marine wildlife.

In a 2020 publication, James and colleagues used a combination of field surveys, biomechanical measurements, and wave modeling simulations to assess how Caribbean seagrass meadows withstand Category 4 and 5 hurricanes.[17] They found that the native seagrasses were resilient to major storms, such as Hurricane Irma (a Category 5 hurricane), and managed to substantially attenuate wave energy, thereby making a valuable contribution to coastal protection against the impact of the storm. With climate change, as increasing sea surface temperatures are making higher-category storms more likely, seagrasses can serve as a natural and sustainable way of mitigating the expected impact of such events.

Historically, seagrasses were often considered a marine weed that crops up and disappears without making much of an impact. However, the closer researchers have looked at the seagrass meadows, the more ecological interactions they have discovered. Numerous animal species, including large ones like sea turtles and the dugong (*Dugong dugon*; vulnerable) feed on the plants. Moreover, like coral reefs, the structured and stable environment created by the plant can serve as habitat for many invertebrates and as nursery space for numerous fish species. But how stable is this structured environment?

Challen Hyman from the University of Florida in Gainesville and colleagues studied more than 50,000 mollusk shells, both from present (living) and past (dead) residents of seafloor sites, with or without seagrass vegetation, along the Florida Gulf coast to assess how the plant environment affects invertebrate biodiversity, and how it changes over time.[18]

Comparing present vegetated sites to those with bare sediment, they found substantial differences in the living mollusk populations, with the vegetated sites supporting much more local biodiversity and better stability across space, meaning that similar seagrass sites supported similar mollusk communities.

Because many mollusk species can be readily identified by their shells, the authors were able to extend the analysis into the past. Radiocarbon dating revealed 40% of the shells collected to be more than 500 years old, with the oldest specimens dated at almost 2,000 years old. Over this timescale, the authors found that sites currently vegetated with seagrasses have remained remarkably stable ecosystems, comparable with the longevity of terrestrial forests. Unsurprisingly, sites with bare sediment did not show this kind of long-term stability.

For conservation, this means that saving the surviving seagrass meadows is preferable to replacing them with new sites. This is also in accordance with climate considerations, given the organic carbon stocks that these sites have bound.

Beyond mollusks, seagrass meadows also support vertebrate diversity. Marine conservation groups focusing on animals like sea turtles are among those that have underlined the importance of seagrasses. But the fishing industry would also be well advised to engage with seagrass conservation.

Seagrass meadows around the world play an important role for fisheries, as they provide nursery habitat for many species, including commercially important ones like the Alaska pollock (*Gadus chalcogrammus*). Due to their nearshore location and easy accessibility, seagrass meadows are also prime locations for artisan, subsistence, and recreational fishing.

Richard Unsworth at the University of Swansea, United Kingdom, and colleagues conducted a worldwide study of the extent,
importance, and nature of the exploitation of seagrass meadows
for use as a fishery resource.[19] They found that people are targeting these ecosystems around the globe for fishery locations,
mostly in an unregulated and undocumented way. These activities are important for the food security of many people in the developing world, but they also create the risk of a "tragedy of the
commons," in that people are endangering the very resources they
depend on.

The authors conclude that there is a general disconnect between
management policies of fisheries and conservation efforts: "Fisheries modeling and management approaches tend not to consider
the functional role of seagrass and other coastal habitats on re-
cruitment to the spawning stock, for example, current U.K. marine protected area policy." Thus, given that seagrasses are
vulnerable to many disruptive factors and declining at an estimated rate of 7% globally, more connected thinking is needed to
protect the role they are playing in global food security.

Unsworth is also the chief scientific officer and a co-founder
of the conservation charity Project Seagrass, which aims to restore lost seagrass meadows by depositing sandbags spiced with
seeds. A pilot project bringing out a million seeds in Dale Bay on
the Welsh coast has recently seen its first green shoots sprouting.
The project also engages in outreach work, visiting schools to raise
awareness of the importance of seagrass for fisheries and the
climate.

A broader umbrella under which seagrasses could gain some
much-needed policy support is the concept of blue carbon, referring to the carbon that is captured and stored by coastal vegetated
ecosystems including seagrass meadows as well as tidal marshes and

mangrove forests. All these previously neglected coastal ecosystems can make quantifiable contributions to carbon sequestration and thus become part of official climate policy.

In 2020, Jeffrey Kelleway at the University of Wollongong, Australia, and colleagues presented an assessment of blue-carbon strategies that Australia could adopt nationwide.[20] Based on their previous finding that Australia hosts between 5% and 11% of the estimated global blue-carbon stocks, the authors argued that their sustainable management could make a sizable contribution to the country's climate effort.

The authors analyzed 12 separate types of policy action for their potential to abate greenhouse gases in a quantifiable way, in order to identify the five most promising. Among the five winners, they find two specifically addressing the conservation of existing seagrass meadows, with one focused on water quality and revegetation, and the other on the protection from physical disturbance.

Kelleway and others were also involved in a global assessment of the outlook for blue carbon led by Peter Macreadie from Deakin University in Geelong, Australia.[21] The authors discuss key questions and challenges that need to be addressed and call for a comprehensive research program dedicated to blue-carbon science to close the knowledge gaps still hindering its optimal use in climate mitigation.

Blue-carbon strategies have the advantage of making a calculable contribution to the mitigation of climate change while offering additional benefits, including coastal protection and support for fisheries as well as for vulnerable species. Even more reasons to afford seagrasses the attention of global conservation efforts on the scale we already lavish on their animal counterparts, the whales and dolphins.

Cetaceans on the brink

It was a desperate last-ditch attempt in a failing conservation effort, and it ended in tragedy. In 2017, a large team of conservation experts captured two vaquita porpoises (*Phocoena sinus*) in a bid to protect them in a sea pen. The first, a juvenile, soon showed signs of stress and was released. The second, an adult female, died of a stress-induced heart attack soon after capture. Tissues of the deceased animal were saved for scientific studies, but the capture plan was swiftly abandoned.

With an estimated population of fewer than 20, the vaquita porpoise is often referred to as the most endangered marine mammal. It is only found in the warm and shallow waters in the northwestern part of the Gulf of California, Mexico, and became famous by its affectionate Spanish name, which is a diminutive of cow (*vaca*), so could be translated as "little cow." With an adult body size up to 150 centimeters (59 in.), it is the smallest cetacean known.

The species was first described in 1958, so there is no record of its abundance over the longer term. Its estimated population of around 600 vaquitas in 1997 shrank to 250 in 2008, raising the alarm for conservation efforts. Gill-net fishing was identified as a major cause of vaquita mortality. Gill nets are open nets deployed as a vertical curtain, which mainly work by entangling fish of a specific size trying to swim through them. Typically, the fish's gills become stuck in the net, which explains the name. Mammals becoming entangled in these nets die by drowning.

Gill-net fishing was banned in the vaquita's habitat from 2015 onward. However, illegal fishing operations targeting the totoaba (*Totoaba macdonaldi*), a large fish that is also endangered, continue to use the type of gill nets that are the most lethal for vaquitas because they are of similar size to the fish targeted. The totoaba is traded for its swim bladder, which is sold on black markets in China.

Illegal gillnetting is the reason for the continuing decline of vaquitas and may well drive them into extinction by the time this book is published. However, a tiny glimmer of hope has emerged from recent studies resulting from the 2017 failed capture attempts.

When one of the captured vaquitas died in the failed 2017 rescue operation, a team of experts was ready to safeguard its tissues and cell lines for future conservation research. With these materials of exceptional quality for a rare and elusive species, Phillip Morin from the National Oceanic and Atmospheric Administration (NOAA) Southwest Fisheries Science Center in La Jolla, California, obtained funding for a team led by Olivier Fedrigo, Jacquelyn Mountcastle, and Erich Jarvis at the Rockefeller University in New York to sequence a reference genome to a quality unmatched for any other cetacean species.

By 2020, fewer than 1% of species listed as threatened had a reference genome published that met the platinum quality standard set by the Vertebrate Genome Project, including only one other cetacean species, the blue whale (*Balaenoptera musculus*).[22] As Annabel Whibley from the University of Auckland, New Zealand, discussed in a perspective accompanying the genome publication, the availability of such a reference-quality genome can transform conservation efforts by providing important genetic details as well as insights into population structure.

Interpreting the genome, Phillip Morin and colleagues were looking out for the demographic history and genetic health of this extremely small population.[23] They found that the level of differences between the maternal and paternal chromosomes of an individual (heterozygosity) was very low, as expected in an inbred population. The heterozygosity value they determined was even the lowest of any mammalian species determined so far. However, they also found that this genetic homogeneity was evenly spread across the entire genome, which was not the pattern expected of inbred individuals.

Therefore, Morin and colleagues concluded that the vaquita population has not suffered a recent bottleneck effect and fallen into an "extinction vortex"—the spiral of decline that many species face when a shrinking population leads to inbreeding depression (reduced fitness). On the contrary, it appears the species has been thriving as a very small population, with an effective population size below 5,000, for many generations, with estimates ranging from 200,000 to 300,000 years. Its lineage split from the common ancestor of its closest relatives—the two porpoise species of the Southern Hemisphere, Burmeister's porpoise (*Phocoena spinipinnis*) and the spectacled porpoise (*P. dioptrica*)—some 2.5 million years ago.

The interpretation of these findings suggests that, by living in a small population for an extended period, the species has been able to purge any deleterious mutations that might have spelled disaster in a sudden shrinkage with a genetic bottleneck effect.

This healthy genome agrees with field observations. As recently as 2019, researchers have been able to observe vaquita mothers with healthy calves. It gives conservationists hope that they may be able to bring the species back from the brink of extinction. They only have to save the surviving individuals from poachers and gill nets.

To that purpose, the Sea Shepherd Conservation Society has been working with the Mexican government to remove illegal gill nets from the protected area. "Operation Milagro" has been running for more than seven years using several vessels of the Sea Shepherd fleet. As of March 2023, an estimated population of 8 to 10 vaquitas is still clinging on, and the fight against illegal gill nets continues.

Before the vaquita genome, Morin's group described similar low genetic diversity in the absence of a genetic bottleneck or extinction threat in other cetacean species, including the globally distributed killer whale (*Orcinus orca*) and the short-finned pilot whale (*Globicephala macrorhynchus*). Killer whales display a complex global distribution: populations at lower latitudes feature low diversity, while Arctic and Antarctic populations tend to be more diverse and divergent from the genomic consensus.[24]

Similarly, Michael Westbury at the Natural History Museum of Denmark in Copenhagen and colleagues have shown that the narwhal (*Monodon monoceros*) maintains a low genetic diversity in spite of its abundant population size and that this is not the result of a recent inbreeding or bottleneck.[25] The researchers conclude that after a long period with a small effective population size, narwhals probably expanded to current levels around the time of the Last Glacial Maximum.

Despite their healthy population size, which has led to the narwhal's Red List status changing from near threatened to least concern, the species is still considered vulnerable to the impacts of climate change. As the narwhal is highly specialized to the Arctic environment, the rapid ice loss and warming of its habitat could still affect its viability. The low genetic variability may turn out to be a disadvantage when the species is forced to adapt to a warmer environment.

Other cetacean species, including some teetering on the brink of extinction or recovering after a brush with death during the era of industrial-scale whaling, may also benefit from advances in genome and population studies. Apart from the vaquita and the Yangtze River dolphin or baiji (*Lipotes vexillifer*), which hasn't been confirmed with physical or photographic evidence since 2000 and may already be extinct, the IUCN lists two other cetaceans as critically endangered—the Atlantic humpback dolphin (*Sousa teuszii*), which is found only off the coast of West Africa, and the North Atlantic right whale (*Eubalaena glacialis*). In addition, Rice's whale (*Balaenoptera ricei*) was described as a new species in 2021 and is already categorized by IUCN as critically endangered.[26]

Right whales, historically described as one species but now split into three, became one of the main casualties of whaling—the name derives from being considered the "right" whale for whalers to target. Of the three separate species now identified, the North Atlantic right whale is the most threatened, with an estimated population of under 400. Collisions with ships and entanglement in fishing gear are leading causes of mortality in the North Atlantic species, which has a markedly shorter life expectancy than the southern right whale (*E. australis*). Its population continues to shrink as mortality consistently exceeds birth rates.

Although conservation groups and marine authorities in the United States and Canada have worked to reduce deaths with measures such as speed restrictions for vessels, death level remains unsustainable. Given the rapid warming of the Arctic leading to loss of sea ice and increased economic activity and ship traffic in the area, the situation could become even worse for this species.

The DNA Zoo consortium, which creates genome assemblies of numerous species in a quality comparable to the results of the Human Genome Project, released assemblies for 16 cetacean species during the years 2019 and 2020, including a version of the *E. glacialis* genome at the end of 2019. A collaboration with the Vertebrate Genome Project to assemble a reference genome is under way, co-ordinated by Morin. Its detailed investigation will help conservation scientists to better understand the population structure, demographic history, and survival chances of this severely threatened species.

Some cetacean species have made remarkable recoveries, coming back from the brink of extinction after the international moratorium on commercial whaling came into force in the mid-1980s. Humpback whales (*Megaptera novaeangliae*), beloved and much studied for their social learning of complex songs (see chapter 8), are a case in point.

Alexandre Zerbini from NOAA's National Marine Fisheries Service in Seattle and colleagues conducted detailed simulations to estimate the decline and recovery of the western South Atlantic population of humpback whales.[27] Using more comprehensive data than a previous assessment, run by the International Whaling Commission (IWC) between 2006 and 2015, of all humpback populations in the Southern Hemisphere, the researchers arrived at higher estimates for the population before exploitation, making for an even deeper decline but also an impressive recovery.

The western South Atlantic population migrate between winter breeding grounds off the Brazilian coast and summer feeding grounds around South Georgia and the South Sandwich Islands. Commercial exploitation began at the beginning of the nineteenth century, but humpbacks in the South Atlantic were most heavily targeted from 1904 onward, when whalers set up a whaling station on the island of South Georgia.

Considering both the whales recorded upon landing and the estimated collateral damage, including whales killed but lost at sea and calves dying as a result of their mother being hunted, Zerbini and colleagues estimate that between 40,000 and 60,000 whales of this population have been killed by humans since the beginning of the nineteenth century.

Based on these death tolls, the model calculations suggest that the population was at around 27,000 in 1830 and remained relatively stable until large-scale commercial whaling in the South Atlantic started in 1905 and expanded into the feeding grounds at higher latitudes. Unsustainably large numbers were caught around South Georgia, in particular. The population collapsed from 25,000 in

1904 to around 700 in 1926. The population remained small for decades, with the lowest point estimated to have been 450 individuals around 1958—less than 2% of its size before exploitation.

Hunting ceased by 1972, and since then the population has recovered to an estimated 25,000 whales, or 93% of the abundance in 1830, according to Zerbini and colleagues, who used a combination of air- and ship-based surveys, along with advanced modeling techniques to assess the current abundance. The earlier IWC assessment had estimated a recovery of only 30% by 2005. The authors predict that the whales will have fully recovered by 2030, as long as there is no increase in anthropogenic disturbances, which in their habitat have remained moderate so far. As in 1830, this abundance will likely be close to the carrying capacity of the habitat, as the availability of the krill that the whales feed on may limit the whales' numbers.

Globally, the species has recovered to an estimated two-thirds of its pre-whaling abundance, leading to its Red List status changing to least concern.

Even though the situation looks desperate for species like the vaquita and the North Atlantic right whale, it is comforting to know that miracles can happen.

CHAPTER 7

Living with Animals

Hunters, as we saw in chapter 5, became more and more efficient at killing all kinds of animals that might serve as food and drove many species into extinction. When humans settled down and started to grow their food, however, a different kind of dynamic developed. Animals, such as rodents, cats, dogs, horses, and donkeys, soon surrounded the early farmers, and each has a different story of an evolving ecological connection with humans. Over the last few years, studies of modern and ancient genomes of many human-associated species have revealed when and how we learned to live with animals.

Since the Industrial Revolution, the human habitat has shifted to cities, which now home most of the world's population. Still, our connections with tamed animals endure even in the urban environment. Moreover, although our concrete-clad cities are a hostile environment for most wildlife, there are still species that have opportunistically made the most of the new habitat, or at least managed to tolerate its dangers, giving rise to a whole new scientific discipline of urban ecology. From bedbugs to parrots, the list of urbanized species is long and may contain surprises. Studies already address how this new environment shapes the evolution of some species.

Of mice and men, cats and grains

Our species has been the most invasive of all, spreading to all continents within the last 70,000 years and exterminating most of the terrestrial megafauna on most continents (see chapter 5). Another major human impact on global biodiversity is the help we provide for other species to invade. This is obvious for the small number of plant and animal species we breed for food, from rice to cattle. The most successful human-assisted invaders, however, are rodents, and among them is the humble house mouse.

Naïvely, one could expect the sequence of events at the dawn of agriculture to be a straight line: people started to grow cereals, so they had to keep grains over winter for sowing in the spring, which attracted mice, which in turn became a target for wildcats. Humans welcomed the cats because they kept the rodents in check. This causality chain might explain how we ended up with pet cats in our houses as well as countless expressions, fables, and nursery rhymes reflecting millennia of cohabitation with mice.

A comprehensive multifaceted analysis of remains of ancient rodents and genomic studies of cat domestication suggest a somewhat more complex story, however. Thomas Cucchi from the Muséum national d'Histoire naturelle in Paris, France, and colleagues set out to clarify the origins of the western house mouse, *Mus musculus*, and analyzed more than 800 dental remains of mice (*Mus* sp.) from the Near East and southern Europe.[1] Two subspecies of the European house mouse are relevant to this story—namely, *Mus musculus domesticus*, currently present in southern Europe, and *Mus musculus musculus*, in central and northern Europe. With European colonialism, both species spread around the world. The researchers had to distinguish these species from superficially similar regional, non-domestic mouse species.

The researchers used geometric morphometrics analyses and, where possible, mitochondrial DNA to identify the species and subspecies. They also used carbon dating and archaeological information about species' connection to human settlements. The analyses proved a challenge as only a fraction of the rodent samples found yielded genetic sequences or a reliable age, but by combining the morphological analyses and the limited molecular data, the researchers eventually gathered enough data points to suggest a course of events.

The *M. m. domesticus* subspecies originated in the Iranian Plateau, where specimens were found in accumulations presumably left by birds of prey. From there it came to Mesopotamia, naturally expanding into the area before the Natufian culture established the first human settlements there. Cucchi and colleagues dated the earliest *domesticus* samples associated with human settlements to 14,500 years old. This clearly places the mouse in a context where hunter-gatherers established a more sedentary lifestyle, but 2,000 years before they started storing grains. The authors suggest that the benefits for the rodents at this stage included shelter from the elements and from predators, as well as the occasional morsel of food. This was only a viable niche in the largest of these pre-agricultural settlements, but it put the house mouse in a prime position when humans started growing cereals and storing large amounts of grain.

The origins of agriculture in the Fertile Crescent of the Near East have been studied in detail. Far from being a human invention that improved life for hunter-gatherers who made the switch, agriculture is best described as a co-evolution of humans and the species they used, including barley in the Fertile Crescent around 10,000 years ago. Unintentional human effects on plant evolution started when people started gathering large proportions of the grains produced and thereby made the natural seed dispersal mechanisms ineffective. Research suggests these gradual and unconscious effects started millennia earlier than previously thought, possibly as early as 30,000 years ago for einkorn and 25,000 years ago for emmer wheat.[2]

This use of wild cereals evolved into planned cultivation of crops, which enabled higher population density, accumulation of wealth, and diversification of tasks. Agricultural societies were able to feed armies and thus conquer land used by more dispersed hunter-gatherer societies, which explains their paradoxical success in spite of the fact that early farmers lived shorter, less healthy lives than their hunter-gatherer ancestors.

For the mice that had already learned to live in the nooks and crannies of human habitations, the advent of cereal farming and grain storage around 2,000 years later brought an abundance of food that would conceivably have led to rodent infestations very quickly. Cucchi and colleagues find the earliest evidence of *domesticus* associated with farming villages at Jerf el Ahmar (Northern Levant) and Netiv Hagdud (Southern Levant), which they determined to be 12,000 years old.[3]

The transfer of mice between settlements and even between the Mediterranean coast and islands such as Cyprus suggests that mice traveled as stowaways when cereals were transported or traded. Genetic evidence from a single molar dated to between 11,100 and 10,600 years ago, backed up by circumstantial evidence, places *domesticus* in the settlement of Klimonas, Cyprus, at this very early time.

Later settlements on the island, including Khirokitia, on the southern coast, and Cape Andreas-Kastros, on the northeastern tip of the Khyrenia Peninsula, revealed large populations of *domesticus*, suggesting the mice had conquered the entire island by

6000 BCE. No wonder Cyprus was a hotspot for the early appearance of the human-associated cat, as we will discuss below.

With the spread of agriculture into Anatolia, *domesticus* also traveled northward, but it had only limited success in Europe. It established a metapopulation on the coasts and islands of the eastern Mediterranean, presumably with ships transporting grain provisions, which is directly confirmed by the house-mouse mandible found in the cargo of the late Bronze Age Uluburun shipwreck off the southern shores of Anatolia. However, much of the house-mouse invasion of Europe came from elsewhere.

Cucchi and colleagues found that, at the end of the Neolithic, the subspecies *Mus musculus musculus* was the first house mouse to spread into continental southern Europe. Its occurrence is documented from late Neolithic/Chalcolithic household deposits from around 6,500 years ago, from tell sites such as Bucşani in southeastern Romania and Vinča-Belo Brdo in Serbia. At Bucşani, the researchers could determine cytochrome b gene sequences for six specimens and a direct radiocarbon dating suggesting an age between 6,627 and 6,413 years.

Various routes have been suggested for the arrival of *musculus* in Europe, but Cucchi and colleagues suggested that, of the alternative models available, "a human dispersal with the advance of agriculture into Europe through the Pontic Steppe north of the Black Sea has been considered the most parsimonious." Intriguingly, this would suggest that *musculus* may have traveled together with Indo-European languages. The issue is complicated, however, by the possibility that *musculus* may have spread without human help for longer than *domesticus*. The authors suggest that further analyses of animal remains from Neolithic settlements, especially in Ukraine, are necessary to resolve this question.

If the first mice moved into human habitations to find a haven from predators, their descendants proliferated after the advent of

agriculture to the extent that they drew wildcats to the settlements. Realizing that rodents were becoming a problem, early
farmers must have tolerated the cats and may have even encouraged them to move in. The early presence of a cat in a Neolithic
child burial on Cyprus, which had no native wildcat population
before farmers arrived, may conceivably be interpreted as evidence
pointing to the conscious decision of early farmers to bring cats to
the island, presumably for rodent control.

To flesh out this plausible story with firm details, the groups of
Eva-Maria Geigl at the Institut Jacques Monod in Paris, France,
and Wim van Neer at the University of Leuven, Belgium, joined
forces and analyzed ancient DNA samples from hundreds of cat
specimens obtained from museums and archaeological sites, covering a time range from 7000 BCE to the nineteenth century.[4] The
project initially proved challenging due to the poor preservation of
DNA in the climatic conditions of North Africa and the Middle
East, but thanks to the dramatic improvements in sequencing methods in the first decade of this century, it eventually took off, and the
researchers were able to piece together the origins of domestic cats.

From mitochondrial DNA, the researchers found two major
maternal lineages accounting for the largest contributions to today's
global population of domestic cats. Both can be traced to the North
African and Southwest Asian subspecies of the wildcat, *Felis silvestris lybica*, which apparently became associated with human
settlements in separate populations in Anatolia, the Levant, and
Egypt. From Anatolia, Neolithic farmers brought this subspecies
to Europe and the eastern Mediterranean.

The ancient Egyptians famously revered cats as godlike beings,
leading to countless representations in images and indeed to cat
mummies that have provided helpful specimens for the genetic
analyses revealing an additional lineage of *F. s. lybica* different
from the Anatolian one.

The researchers found that Egyptian cats were spread along the marine trade routes of the classical era, which suggests that they were used for pest control on board trading ships. Viking ships carried cats of Egyptian ancestry as far as Northern Germany. While the native European wildcat, *F. s. silvestris* occasionally hybridized with domestic cats, *F. s. lybica* remained the dominant maternal lineage. During the time of the Roman Empire, the Asian wildcat, *F. s. ornata*, or hybrids descended from it, also left genetic traces, obviously traveling with sea traders from India. Thus the genetics of animals like the cat offer complementary information to the human genetic studies that have recently yielded insights into the Bronze Age populations and migrations in the eastern Mediterranean.

Unlike dogs, however, which have been profoundly altered by breeding and the need to fit into human society, cats didn't change that much when compared to their wild ancestors, and they have kept their independent spirit. The genetic analyses confirm that for the longest part of the shared history of humans and house cats, the deal was one of mutual tolerance, rather than active domestication. Thus, the genes show that the coat patterns of the cats in the ancient Mediterranean were gray stripes like those of the wildcat ancestor. Different-looking cats were only bred in Europe from the Middle Ages onward, so human interference starts much later than it did for dogs.

With the age of sailing ships, where rodents were a constant threat to food stocks, cats became an important part of the crew, and thus they spread around the globe with European travelers.

On the origins of donkeys

Apart from cats and dogs, humans also tend to worship horses (*Equus ferus caballus*) in many irrational ways, betting on which horse will run fastest, keeping them at great expense for the pursuit of archaic hunting traditions (or for the sole purpose of sitting on a high horse and looking down on those who don't), and even creating paintings and statues of their leaders on horseback.

By contrast, another closely related species from the same genus, the domestic donkey (*Equus asinus* or *Equus africanus asinus*, depending on whether it is classified as a species or a subspecies of the wild ass, *Equus africanus*) gets barely noticed and sometimes ridiculed rather than admired, even though it has been working for us for a lot longer than horses, carrying loads and pulling chariots even in ancient Mesopotamia. The Bible mentions Jesus riding on a donkey, fulfilling a prophecy that simultaneously marks him as the promised king and as an ally of the common people. This is just one of 20 cameo appearances that the traditional beast of burden makes in Christianity's holy scriptures.

Donkeys also appear in ancient hieroglyphs, Aesop's fables, the Qur'an, *Don Quixote*, and in numerous other cultural products through to children's books and nursery rhymes. While children relate to the long-eared cuteness of the animal, grown-up perceptions often highlight negative traits, such as stubborn fearfulness in the saying "lions led by donkeys."

Although industrialized countries have largely replaced donkeys with diesel engines, the animals are still indispensable as cheap and robust load carriers in many poorer countries, especially in hot and arid climates and in places where transportation infrastructure is insufficient.

As A. A. Milne's creation Eeyore would surely have anticipated, donkeys were neglected by evolutionary genomics for two decades, while every other species of plant and animal that is either useful to or otherwise associated with humans has been studied comprehensively. Genomic studies of barley, for instance, have enlightened us on the origins of agriculture, while studies of ancient cats and mice have shown up the curious history of how this predator-prey couple followed humans and their cereals around the world (as discussed earlier). All those grains and the products made from them needed transporting, obviously, but it was only in 2022 that donkeys finally got their moment in the spotlight as their origin story was revealed.

Domestic donkeys were already known to the earliest cultures of which we have written records, so we can be sure that they have accompanied us and carried our loads for more than 5,000 years. Due to a shortage of archaeological evidence and the sporadic nature of any sequencing efforts, however, the place and time of their domestication has been the subject of competing theories and speculations.

Ludovic Orlando's group at the University of Toulouse, France, and numerous colleagues from institutions elsewhere in France and around the world addressed this problem with additional genomes of under-studied populations and a comprehensive analysis of all available information on donkey genomes since the dawn of donkey history.[5]

The phylogenetic analyses of 49 newly sequenced and 158 publicly available modern donkey genomes were compatible with two conflicting hypotheses entertained so far: a unique origin in Africa and a dual origin in separate locations in Africa and the Arabian Peninsula. Further modeling of the population development favored the out-of-Africa version, however. Thus, based on modern donkey genomes alone, the authors concluded that their "analyses support an early domestication in Africa, spreading at an even rate into the Arabian Peninsula and Eurasia, and flow back into Nubia and Maghreb." The authors calculated that the initial domestication must have happened more than 7,000 years ago, around the time the Sahara became a desert. Among modern populations, the donkeys found in the Horn of Africa and in Kenya most closely resemble the common ancestors shortly after first domestication.

Studying 31 ancient donkey genomes from 11 archaeological sites and covering the last 4,000 years, the authors obtained further confirmation of the unique origin model as well as unexpected new details of the population history. The three oldest samples came from a site in Anatolia (Acemhöyük, Türkiye) and were carbon-dated to 2564–2039 BCE. This dating suggests, in agreement with the phylogenetic studies, that the expansion out of Africa occurred before 2500 BCE.

The ancient European donkeys studied were found to be in line with modern European breeds, except for one specimen from Marseille, which appears to have had origins in West Africa. More generally, however, the authors found that trade links across the Mediterranean changed African donkeys more than European ones.

Further complexities emerged in Iran, where the authors found surprising differences that may be explained by a wholesale population turnover between 1000 BCE and 500 CE. An ancient genome from Israel dated to 350–58 BCE showed the largest influx of genes from wild equines, a heritage it passed on to some populations in Asia but not in Europe. The authors noted that both mitochondrial and Y-chromosome studies had previously failed to pick up the unique position of this ancient specimen.

When the domesticated horse expanded from its origins in Asia to reach Mesopotamia some 4,000 years ago and then Europe, people could choose between equid domesticates and produce hybrids. They soon got the impression that the mule, a first-generation hybrid with a donkey father, is a more patient and useful work animal than either of its parent species. As is to be expected for hybrids formed from separate species, mules are typically infertile (although some exceptions have been confirmed and studied), so the production of mules required specific and continuous breeding efforts with the management of the suitable donkeys and mares.

As part of their investigation into ancient donkey genomes, Orlando's team may have uncovered a likely mule breeding center in the Roman Empire, located in what now is the northeast of France. The researchers analyzed the genomes of three female and six male donkeys from the archaeological site of Boinville-en-Woëvre (in the Meuse department). The nine individuals found in the farming area of a Roman villa turned out to be highly inbred and closely related, although there was also evidence of some outgroup admixture. All of this suggests that they were being bred for specific properties, such as their remarkable height.

As the market in the Roman Empire generally called for mules, which are accordingly widespread in archaeological finds, the authors concluded that the residents of the Boinville-en-Woëvre site did not breed donkeys for donkeys' sake but rather for their use in breeding mules of very specific qualities. According to Orlando and colleagues, "this suggests Boinville-en-Woëvre as a likely mule production center that maintained the bloodlines of giant donkeys selected through familial breeding and restocking."

The authors also noted that the absence of mules and rarity of horse remains at the site suggested a modus operandi that focused on the donkey side of the mule production, offering impregnation services to owners of horse mares either when they are brought in or on visits to their home farms. The ubiquitous use of mules across the Roman Empire would have supported a supply chain with such highly specialized businesses.

Further insights into breeding strategies across the ages gleaned from the genomes suggest that the level of inbreeding stayed roughly constant in donkeys. This contrasts with horses, where the amount of inbreeding increased in recent centuries, as Orlando's group established in a comprehensive investigation into their origins and expansion from the Eurasian steppe around 2000 BCE.[6]

The authors also noted changes in the amounts of outgroup admixture, which decreased over time, but this finding doesn't necessarily reflect breeders' choices. Ancient donkeys may just have lived more freely and thus had more opportunities to sample the wild gene pool.

Sometimes people did cross donkeys with wild animals on purpose, however. The earliest recorded hybrid of this kind is the kunga, which was often mentioned in cuneiform texts as an animal of high value, costing six times the price of a donkey. It is also shown in ancient images, such as the "Standard of Ur," a 4,500-year-old Sumerian mosaic. This equid was associated with Mesopotamian elites, offered to foreign leaders as a particularly valuable gift, and used to pull four-wheeled chariots for travel and war in the times before domestic horses were known in the area.

The exact biological identity of this coveted animal breed had remained controversial until Andrew Bennett from the Institut Jacques Monod in Paris, France, and colleagues presented ancient DNA analyses that may resolve the issue.[7] The researchers studied the equid remains from the 4,500-year-old princely burial complex of Umm el-Marra, located just east of Aleppo in northern Syria. Within the large and richly ornamented site, 25 male equids were buried in separate installations. At least half of them appear to have been killed specifically for the purpose of the burial rites. Based on morphological and archaeological criteria, they are thought to represent specimens of the prestigious kungas.

Although the DNA from the burial was generally too degraded for high-quality genomes, the researchers were able to sequence a partial genome of one of these approximately 4,500-year-old equids, together with genomes of an approximately 11,000-year-old Syrian wild ass or hemippe (*Equus hemionus hemippus*) from Göbekli Tepe in today's Türkiye and two of the last hemippes that were conserved in museum collections before the subspecies became extinct in the early twentieth century. They also used PCR technology to target regions on the mitochondrial genome and Y chromosome of the Umm el-Marra relic that might yield specific information on male and female ancestry.

All genome results indicated that the Umm el-Marra equid was a mixture of equal parts of domestic donkey and hemippe. More precisely, the sex-specific PCR methods show that the animal was a first-generation cross between a female donkey and a male hemippe. The data also suggests that the hemippes at the time of the breeding, with a shoulder height of 130 centimeters (51 in.), were much taller than the last survivors of the subspecies observed in the nineteenth century, which stood at just 100 centimeters (39 in.).

This finding represents the earliest documented example of humans deliberately producing a hybrid species. This would have taken considerable effort considering the hemippes had to be captured from the wild. The reward was an animal of considerable size and strength, suitable for highly valued roles that were later taken over by horses and mules. Based on skeleton comparisons to other wild equids from the family of hemiones, to which the hemippe belonged, the kunga may well have run faster than horses.

Throughout their 7,000-year history in the service of humanity, donkeys typically were the cheapest option, often replaced by horses or mules when these became available and affordable, and continuing to carry our burdens only where there is no better option.

One may wonder, therefore, if the continuing global spread of motor vehicles, apart from tipping our climate and biodiversity over the edge, may also lead to the donkey becoming a conservation concern. You don't have to be Eeyore to be gloomy about their future. Animal rights charities in the industrialized world are already worried about the treatment the animals are experiencing in poor countries, where they are typically getting too little food and too much work to do. Those for whom a donkey was a last resort will not show much mercy on the beast of burden once they can afford a tractor or a lorry and have enough roads to drive it on.

Apart from the small number of animals serving to entertain children and tourists from wealthier countries, donkeys may be at threat of being bred as meat or for traditional Chinese medicine. Some populations may also return to the wild, like California's "burros" did when they were abandoned after the nineteenth-century gold rush. The burros, as well as the Asinara donkey in Sardinia, are now enjoying protected status. Others may need our protection too. After 7,000 years of carrying our loads, the much-maligned donkey could expect humans to show some gratitude.

Adapting to life in the city

In the first half of the twentieth century, most humans lived in the countryside, working the fields to provide food for themselves and city dwellers, who were in the minority. With the globalization of trade and industrial production toward the end of the century, this balance started to shift dramatically and is now leaning to the other side. By 2014, more than half the world population was living in cities, and this proportion is forecast to head toward two-thirds by the midpoint of this century. Within a human life span, we have thus turned from a mostly rural to a mostly urban species.

As cities spread and grow, they and the infrastructure they require for their transportation and energy needs destroy vast amounts of wildlife habitat. Karen Seto from Yale University and colleagues have forecast the urban land coverage to increase by 285% between 2000 and 2030.[8] Accompanied by explosive growth of the road network and motor vehicles, this development constitutes a major threat to terrestrial biodiversity and to the ecosystem services that we all depend on.

In addition to the obvious effects of land-use change and habitat loss, however, the human-made environments are creating new niches for wildlife to exploit and adapt to. Global urbanization provides biologists with unique opportunities to study urban ecology and urban evolution in times of rapid change, and to learn how nature responds and how cities might become more sustainable.

Some species have co-evolved with human settlements from the beginning of agriculture some 12,000 years ago. As we discussed above, the storage of grain and the disposal of food waste opened ecological niches for rodents, including the house mouse (*Mus musculus*), the black rat (*Rattus rattus*), and the brown rat (*Rattus norvegicus*). Following their trail, the house cat moved in and likely domesticated itself. Arthropods like the German cockroach (*Blattella germanica*) and bed bugs (*Cimex lectularius*) also benefited from human settlements and became ubiquitous in towns and cities, only limited by pest control measures.

While these species evolved over millennia on a timescale comparable to the transition of human lifestyles and the domestication of farm animals, the current rapid change toward global urbanization raises the question of how fast nature can cope with the spread of cities, and how quickly species can adapt to urban life, either by behavioral learning or by evolution.

Pigeons (*Columba livia*), which have lived in cities for centuries, are an exceptional case, as they were originally domesticated for food use and then spread as feral urban populations thanks to the availability of nesting sites and food. The twentieth century, with its sprawling suburbs, has seen many other species getting a taste of city life, such as the North American raccoons (*Procyon lotor*) that invaded Europe after an ill-advised introduction in 1934 and foxes. There have been reports of urban wolves in German suburbs, wild boars in Rome, and cougars in Los Angeles. These species are more widespread in cities than the human inhabitants realize, as they can quickly grasp the geography of the space and the timetable of human behavior and learn to remain invisible, avoid the rush hour, or exploit waste disposal systems.

A large part of urban ecology is to be found indoors, however. Although many humans may prefer not to think about this, there are thousands of microbial species and hundreds of arthropod species living in our own homes. "In places like New York City, the indoors now dwarfs the outdoors," explains Rob Dunn from North Carolina State University. "If we want to understand where species are evolving most quickly and even where new species are emerging, our bedrooms, bathrooms and boardrooms are a kind of modern Galapagos. They present a chance to see evolution in action without ever leaving home."[9] In a survey of 50 homes in the United States, Misha Leong from the California Academy of Sciences in San Francisco, with Dunn and others, studied the domestic arthropod biodiversity to established what factors may influence it.[10]

The authors found that the biodiversity varies with the type and location of rooms, finding characteristic differences between basements and attics, as well as between kitchens and bedrooms. Surprisingly, the behavior of the human residents, including whether they keep pets and how often and thoroughly they clean their house, did not make a mark on the level of biodiversity. The factor correlating most strongly with the overall biodiversity found inside a house is the permeability for outdoor species. Therefore, the authors suggested that houses can act like Malaise traps (tentlike textile funnels that entomologists use to collect flying insects) in that they trap a variety of common arthropods from their surroundings, which then move in with the smaller number of more specialized house species, such as cockroaches.

Studies of urban ecology have already been conducted for decades, showing how the human-made landscape of cities can help some species while displacing others, thereby creating entirely new urban ecosystems that may resemble each other more than the rural ecosystems typical of their geographic area.

Fragmentation, hard surfaces, excess heat, and light and noise pollution, as well as chemical pollution of ground, air, and water, are the stress factors that city-dwelling wildlife, much like the human residents, must put up with. A big question is, Does evolution help wildlife adapt to city life? Does the brutal selection in the urban environment bring about new species that are city dependent? Traditional thinking had it that these changes were too fast for evolutionary responses, but there is a rapidly growing body of evidence showing that urban wildlife is indeed evolving as we watch.

One classic example of evolutionary change in response to anthropogenic disturbance is the case of the British peppered moth (*Biston betularia*), which in the nineteenth century changed from light to dark gray in color in line with the building walls in heavily polluted cities. In a recent study of this well-characterized example, Ilik Saccheri's group at the University of Liverpool, United Kingdom, showed that the color change was enabled by a transposable element that emerged in the early 1800s and quickly spread to much of the urban moth population.[11]

Another recent study showing rapid, adaptive evolution in human-made environments was conducted by Elizabeth Kern and Brian Langerhans at the North Carolina State University in Raleigh.[12] These authors studied two fish species from 25 streams in North Carolina, which were classified as rural or urban streams.

Due to the impact of impenetrable surfaces in urban land cover, urban streams generally have faster flow rates and suffer flashes with extremely fast flow rates in the event of heavy rain.

Measurements on nearly 700 individuals showed marked differences between the rural and urban members of the same species. In the case of the blacknose dace (*Rhinichthys obtusus*), the differences in shape were in line with expectations. The urban specimens showed a more streamlined body shape, enabling them to move more efficiently in faster flowing and more unpredictable waters. This finding agrees with previously analyzed differences between fish living in fast- and slow-moving water.

The other species, the creek chub (*Semotilus atromaculatus*), by contrast, showed an equally pronounced but different kind of morphological change in urban waters, which the authors find more difficult to align with theoretical predictions. These fish exhibit a deeper and longer midbody region, which enables them to develop better efficiency during steady swimming.

In addition to the binary comparison between urban and rural streams, the authors also considered the time axis by considering how long a given stream had been urbanized. With this parameter, they could demonstrate that there is an adaptive transition. With additional laboratory experiments exposing rural-born fish to urban-style conditions, the researchers could further demonstrate that the change is not an immediate phenotype response. From this, Kern and Langerhans concluded that the observed differences are likely to have a genetic basis favored by adaptive evolution.

Panagiotis Theodorou from the University Halle-Wittenberg, Germany, and colleagues have used genome analysis to look for effects of urbanization on the red-tailed bumblebee (*Bombus lapidarius*) from nine German cities as compared to those from nine paired rural sites.[13] The researchers found no genome-wide effects but were able to identify several candidate loci that appear to correlate with environmental parameters such as urban land use.

City life has its pros and cons, for bumblebees as much as for everybody else. "On the one hand, food is abundant for the insects thanks to the numerous urban gardens and balconies. But on the other hand, bumblebees have more parasites and there is a considerably higher degree of habitat fragmentation in cities," Theodorou said.[14]

While observational data on urban ecology is amply available and studies showing specific evolutionary changes are beginning to accumulate, biologists have hardly even begun to gain a comprehensive understanding of how the unprecedented global urbanization is changing life on Earth.

To characterize the general nature of phenotypic change and look for connections with urbanization, Marina Alberti at the University of Washington in Seattle and colleagues have recently conducted a meta-analysis of 1,600 phenotypic changes reported from species in various regions. The authors based their investigation on the hypothesis that "shifts in the physical and socioeconomic structure and function of large urban complexes can drive rapid evolution of many species that play important roles in communities and ecosystems."[15]

Alberti and colleagues used quantitative measures of change normalized to the natural variability and expressed using the Haldane numerator, a metric introduced by J. B. S. Haldane in 1949. They found that overall there is a very clear urban signature in the changes reported, but some kinds of anthropogenic disturbances have larger effects than others. The largest impact arises from the dynamic of socio-ecological interactions (i.e., the extent to which cities change the interactions between people within the city and between distant cities) between humans and other species.

Another urban disturbance that ranks highly in the authors' analysis is species introduction. Where humans move in, they invariably bring not only their pets but also their less appreciated companions from rats to pigeons. This disturbance, Alberti and colleagues suggested, triggers more pronounced adaptive responses in the local wildlife than the actual habitat change.

The authors concluded that the pronounced ecological and evolutionary changes driven by urban development are likely to affect ecosystem function and thus also the sustainability of the human settlements that triggered them. A better understanding of these connections will be necessary to understand and predict the impact of the ongoing wave of global urbanization.

In a 2017 review of evolution in urban environments, Marc Johnson from the University of Toronto, Canada, and Jason Munshi-South from Fordham University in New York analyzed the existing evidence linking genetic change—including both adaptive evolution and neutral gene flow—to city environments.[16]

The general fragmentation of populations in the urban environment restricts the gene flow and increases random genetic drift, leading to a loss of genetic diversity within isolated populations and increasing differences between them. This phenomenon has been studied in white-footed mice (*Peromyscus leucopus*) in New York City, where previously connected populations became isolated in the city's parks. When species move into newly built urban areas, there can also be pronounced founder effects (i.e., genetic bottlenecks due to the limited size of the founding population). Red foxes (*Vulpes vulpes*), which only recently colonized the city of Zurich, Switzerland, show such founder effects.

Isolation effects have also been described for mosquito populations independently colonizing underground systems in different cities, and for the yellow-necked mouse (*Apodemus flavicollis*) in Warsaw, Poland.

Johnson and Munshi-South noted that studies of adaptive evolution in urban environments are still too few and unsystematic. Apart from the British peppered moth mentioned above, they highlighted studies of house finches (*Carpodacus mexicanus*) that evolved different beak shapes in the city of Tucson, Arizona, because they depend on sunflower seeds from bird feeders. Several examples of plant and animal species evolving resistance to specific pollutants have also been reported, including killifish tolerating polychlorinated biphenyls (PCBs).

Tales from the urban jungle

In the ancient fable attributed to Aesop, the town mouse and the country mouse disagree on which location offers the better life. The countryside has less food diversity, but the city life is fraught with dangers. The story has been reworked countless times by authors from Jean de La Fontaine (1621–1695) to Beatrix Potter (1866–1943), and similar tales exist in non-European cultures.

Whereas the fable uses the animals as ciphers to stand in for humans with different affinities for country or city life, the same choice also applies for real rodents and many other animal species, especially now as rapid urbanization eats away rural habitats. Like Aesop's town mouse, some animals and indeed plants may find city life preferable, while others can't handle it. Recent research is now providing a better understanding of the continuum between these extremes, which can be studied in groups of related species such as bats.

In an effort to develop widely applicable assessment methods for the urban affinity of species, Janis Wolf from the Freie Universität Berlin and colleagues at other institutions in Germany conducted a metastudy of 356 bat species from around the world, roughly a quarter of all bat species known to science.[17] The researchers analyzed the traits and habitat preferences reported for these species using several different methodological approaches to quantify urban affinity. Using geo-referenced occurrence records in combination with different proxies for urbanization, the authors tried eight different indices for urban affinity.

The researchers found that the simpler indices worked as well as the more complex ones. Although the application of different criteria resulted in different rankings of the species on the urban affinity scale, these differences did not affect the association of species traits with the habitat preference. Therefore, the authors suggest that the simpler methods could also be useful for assessment of other groups of species and help researchers determine the traits favored in urban environments.

The specific traits found in urban bats include low frequencies in their echolocation calls, longer call durations, small body size, and a higher flexibility in their choice of roost locations. This allows bats to switch sites when they are disturbed by human activities.

In bats and in other species to which similar methods could be applied, knowing which traits do and don't favor survival in an urbanized context will enable researchers to predict which species will require conservation attention as their habitats are encroached on by human settlements.

Apart from having innate traits that favor urban living, many animals can also improve their chances in the city by cultural learning. Anecdotal examples include the spread of the ability to open milk bottle caps to get to the cream, observed among blue tits in England a century ago.

A recent study demonstrates this kind of technology transfer among individuals of a species in the urban populations of wild sulfur-crested cockatoos (*Cacatua galerita*) in Sydney, Australia.[18] Sydney-based biologist Richard Major from the Australian Museum had observed how the birds pried open the hinged lid of a garbage can with their beaks and then shuffled along the edge to lift the lid further and tip it over. Major alerted behavior scientist Lucy Aplin, who was then at Oxford University, United Kingdom, but moved to the Max Planck Institute of Animal Behavior in Konstanz, Germany. Based on an online survey launched in 2018, Aplin, with Barbara Klump and colleagues in Konstanz and Sydney, established that the behavior spread from 3 suburbs to 44 in less than two years. Moreover, it became clear that it spread locally, from one area to others nearby. The researchers concluded that birds learned the skill by watching others. After marking birds to find out which individuals acquired the skill, the researchers found that, in a population where the ability was present, only 10% of individuals could tip over the lid, most of whom were males. Individuals of higher rank in the hierarchy were also more likely to learn the technique. Once a bird had opened a bin, however, the others were also able to take part in the scavenging.

In different suburbs, residents observed subtly different techniques, adding to the evidence that knowledge spreads locally and evolves to form regional subcultures. Only one bird seems to have reinvented the methodology spontaneously, starting a new spread in a suburb far from existing schools of bin-opening.

As an intellectual achievement from a member of the parrot order, this isn't all that surprising. Many parrot species are highly intelligent, and a few have established feral populations in other cities too. Back in 2009, one cockatoo of the same species, known as Snowball, enchanted the world with his dance moves, which were experimentally confirmed to be synchronized to a musical beat. It is remarkable to be able to watch as a new skill spreads through a wild population adapting to city life and learning to make the most of the opportunities on offer. One can infer that many such learning processes must be happening in less visible locations, offering new opportunities to a wide range of urban-dwelling species, from rats navigating sewers to birds nesting in artificial structures. Thus, by studying the more visible cockatoos, the researchers hope to generate a broader understanding of how animals learn to live in cities.

Some species bring innate abilities that are useful for urbanized environments and others learn to reap the benefits of the city life. The big question is whether and how urbanization affects evolution. While the observation of ecological niches and urban-affine residents has a long tradition, the understanding of urbanization as a large-scale human-made experiment in evolution is a more recent development, as discussed in the previous section. Some studies have looked at local or regional rural to urban gradients and their effects on evolution. A major study published in 2022 has for the first time demonstrated parallel urban evolution in multiple cities around the globe.

In the Global Urban Evolution Project, led by Marc Johnson at the University of Toronto, Canada, researchers from 160 cities on six continents collected more than 110,000 samples of the cosmopolitan plant white clover (*Trifolium repens*) in locations chosen along gradients from rural to urban environments.[19]

Specifically, the researchers analyzed the clover's ability to produce hydrogen cyanide (HCN) when cells are injured. This is a defense against herbivores, and the mechanism is known to depend on specific gene variants in two separate places in the genome. Inactive variants are present globally, offering the variation for natural selection to work with. Other stress factors such as frost and drought are also known to affect the cost-benefit ratio of HCN production. The working hypothesis was that the combined influence of these factors could drive an adaptive change in the use of HCN for defense depending on the degree of urbanization.

The researchers concluded from their findings that white clover can rapidly adapt to urbanization in locations around the world, although the strength and direction of the response varies between sites. In addition to the known drivers, including exposure to herbivores, drought, and frost, the authors noted that other factors, such as pollution or gene flow from agricultural varieties, may add to the complexity. Mechanisms of such additional influences were yet to be explored.

A better understanding of the rapid and global adaptive processes presumed to happen in other species might help with a range of issues, from conservation to pest control, and thus also serve human well-being, the authors noted.

In a separate, smaller study, Ruth Rivkin and Marc Johnson studied a local example of urban evolutionary ecology in their city of Toronto. They sampled *Impatiens capensis*, a native wildflower important for native bees and hummingbirds, from locations in and around Toronto.[20] Sequencing DNA from 43 parent-offspring pairings, they analyzed genetic diversity and population structure. They found that the plants from more urban environments showed reduced genetic diversity, making them more vulnerable to long-term decline.

Dangers and stress factors for plants and animals living in urban environments typically arise from human activities, such as vehicle traffic. In addition, the built environment replacing vegetation and natural surfaces can affect the suitability of an urbanized area as habitat. The impacts of structural differences and ongoing human activities can be difficult to separate, as the most altered areas tend to be the busiest too. The COVID-19 lockdowns, however, by dramatically reducing human movements in otherwise unaltered cities, offered a unique opportunity to study the effects of human activity separately from those of the built environment. There were many observations of wildlife exploring the deserted city spaces.

In March 2020, as major cities across Europe went into lockdown and the global economy almost came to a halt, many observed that the disturbing, apocalyptic scenario had its upsides. The usually turbid water in the canals of Venice cleared. Neighborhoods normally humming with traffic noise got to hear bird song instead. Wild boars were seen ambling on Las Ramblas in Barcelona, goats took over the Welsh seaside town of Llandudno, and foxes got more adventurous on the streets of London.

Meanwhile, Olivia Sanderfoot from the University of Washington in Seattle and colleagues started a citizen science project using the eBird platform and aiming to monitor birds in cities in the Pacific Northwest of the United States during a time that included lockdown periods.[21] Whereas other field observation studies were forced to pause due to mobility restrictions, the format of a community project where participants submit observations from a space they can safely monitor, such as their backyards or a nearby park, offered the opportunity to study urban ecology in a unique situation of suspended human activity, which the authors refer to as the "Anthropause."

More than 900 volunteers committed to providing at least one survey per week during a three-month period from April to June 2020. Many were enthusiastic because it gave them something meaningful to do while their normal activity was interrupted. In combination with additional data on land cover and traffic movements, the community observations enabled the researchers to study how human activities affect habitat use of 46 bird species present in cities.

The researchers found that three-quarters of the species monitored changed their habitat use when human activity levels changed, including iconic species of the region such as black-capped chickadees, great blue herons, downy woodpeckers, and Wilson's warblers. Intriguingly, however, songbirds did not become more widely audible when traffic ceased, as one might have expected. The authors concluded that at least some songbird species are adjusting their output to remain audible in the presence of anthropogenic noise pollution.

Other findings suggest that, in response to increased human activity, birds are more likely to visit the quieter green spaces that many participants used for their observations, such as parks and gardens. This was observed for the majority among the 23 species that were more often observed during intense human activity, suggesting that they use green spaces as refugia. On the other hand, those whose observability increased with reduced human activity did not particularly seek out green spaces, suggesting that they took advantage of the reduced human disturbance by using a wider spectrum of habitat.

As a one-off citizen science project in response to the lockdowns, the study had obvious limitations, such as the fact that the observation sites were self-selected by the participants. Therefore, the sampling was not representative, as the authors admitted. Differences between the participants in terms of experience and expertise may also have been a confounding factor. The very nature of the unique situation means that there may not be an opportunity to repeat or reproduce the observations. It could serve as a model, however, for studies accompanying smaller-scale changes, such as temporary road closures or traffic-calming measures.

The Anthropause may remain an exception, but the Anthropocene is moving onward, with human-made changes affecting virtually all ecosystems on our planets one way or the other. Urbanization is progressing particularly fast in the developing world, where new megacities are often growing uncontrollably and with limited attention to sustainability issues.

Some thought is being spent on how cities of the future are to be made sustainable, with respect to their environmental impact, as well as livable for their human residents. In this context, a better understanding of how the built environment accommodates and affects wildlife will be crucial.

Observations of city life have a long tradition and can be readily expanded with the help of the human residents. A detailed understanding of the incipient evolutionary change happening in the urban jungle is only just beginning to be sought and will be harder to attain. In a roadmap published in 2019, Ruth Rivkin and colleagues formulated six questions to guide research, addressing how urbanization affects (1) non-adaptive evolutionary processes; (2) natural selection; (3) convergent evolution; (4) the influence of environmental heterogeneity on evolution; (5) the roles of plasticity, ancestral traits, and contemporary adaptation for the ecological success of urban species; and (6) the evolutionary diversification of novel traits, genes, and species.[22]

The authors call for the insights gained to be applied in city planning, conservation, pest management, and public engagement. The hope is that application of deeper understanding will reduce the impact of urbanization on biodiversity and make cities more livable for humans as well as for other species.

A culture of coexistence

Once upon a time, a princess in India gave birth to a son out of wedlock. According to many myths and legends, the child had a habit of turning into a tiger whenever his mother wasn't looking. As a tiger, he went about and killed local people's livestock, leading disgruntled farmers to pick up arms, plotting to kill the feline. The mother negotiated a peace deal that involved her tiger son moving to the forest and the farmers in return installing a shrine for the *wagh* (Marathi word for big cats) deity, Waghoba, and holding annual festivities with offerings of chicken and goat meat.

Many origin stories along these lines are told in large parts of India where people live in the vicinity of leopards or tigers. The culture of Waghoba is still observed with annual rituals in many populations, often on the first day of Diwali or on a specific day shortly after. These cultural traditions go back many centuries, but only recently has conservation science started to appreciate them as an established way of fostering peaceful coexistence with dangerous animals.

Ramya Nair from the Wildlife Conservation Society–India in Bengaluru and colleagues studied the Waghoba belief system and traditions among the Warli people in the state of Maharashtra, on the west coast of India.[23] The researchers focused their study on the districts Mumbai Suburban, Palghar, and Thane, with a range of environments from urbanized settlements to several protected areas, including Sanjay Gandhi National Park, Tungareshwar Wildlife Sanctuary, and Tansa Wildlife Sanctuary. The area is home to leopards as well as jungle cats (*Felis chaus*) and has historically also hosted tigers.

Using the methodology of ethnography, but with conservation implications in mind, Nair and colleagues conducted fieldwork over six months, which involved visiting and documenting Waghoba shrines, interviewing the local Warli people, and observing local customs including the rituals.

As part of the research, Nair attended the annual Waghoba festival, Waghbaras, at three different sites, including one in rural and two in semi-urban surroundings. "These ceremonies typically last for two days and the intervening night, with traditional music and dancing throughout the night. Relatives and friends of participants from neighboring villages also attend the ceremony. Members of the community, even those that have moved away to other parts of the world[,] return at this time to participate in the annual worship rituals for Waghoba," she reported in the paper. "The shaman led the ceremony and performed all the rituals with the remaining participants following his directive. People offered a variety of things as per their ability, from flowers, coconuts, and incense to toddy (fermented palm drink), chickens[, and] goats. The idols are also smeared with vermillion paste, which is considered auspicious. Orally passed down chants and songs dedicated to Waghoba were presented throughout the festival days and nights."

The researchers documented over 150 shrines, finding one in every Warli village they visited and often an additional one on a nearby hill. Speaking to locals about their belief system, the authors established that people identify any large feline wildlife they encounter with Waghoba and believe that they are protected by a mutual agreement with the deity. One participant is quoted as saying, "The wagh is known and accepted as the king of the jungle. We pray to him so that he protects us and does not do us any harm." When the animals do attack livestock or humans, they often seek explanations in their own failure to adhere to the rituals required.

While it is easy to dismiss this belief as superstition (not to mention victim blaming), the authors emphasize the importance of the fact that it does assign a role to the animals in the shared environment. Rather than competing for space and resources, humans and felines are each expected to play their part to keep the peace.

This also means that during an encounter with the animals, people will neither attack nor panic and run away. One story related to the researchers reported a man telling the animal, "If you are going to eat me then go ahead. You are our god." The man closed his eyes and prepared to die, but the cat just walked away. Conceivably, by helping people to stay calm when confronted with leopards or even tigers, the belief may save lives.

The protective effect for the animals is even more straightforward. Being identified with and worshipped as a deity is probably the best protection an animal species can get. Still, the authors felt that conservation science and policy don't sufficiently appreciate this type of peaceful coexistence established over many centuries.

International organizations led by Western countries may come into an area with preconceived ideas about how the local wildlife is to be protected. Therefore, the study aimed "to diversify the way we understand and approach human-wildlife interactions" Nair said. "Something we have tried to speak about through this paper is the idea of coexistence, not necessarily peaceful and devoid of fear, but as something that is constantly negotiated since negative instances of human-wildlife interactions have existed (and found a symbolic place in myths and narratives) since generations for the Warli. Relations between the Warli and Waghoba (both the deity and animal) are multifaceted, which we believe allows people to be accepting of them and coexist with them, despite negative consequences occurring."[24]

The authors suggest that conservation organizations, rather than imposing prefabricated solutions, should engage with the knowledge that local cultures have about wildlife, and with the established role of shamans as negotiators between the interests of residents and animals.

The much more negative attitude toward those predators that don't have an established cultural role became obvious in a separate study on the dhole (*Cuon alpinus*), an endangered canid native to the Himalayas, published together with the Waghoba work as part of a special issue on coexistence with wildlife. Interviewing residents in Bhutan, Phuntsho Thinley from the Ugyen Wangchuck Institute for Conservation and Environmental Research in Bumthang, Bhutan, and colleagues found that a majority of people interviewed opposed conservation measures for the species.[25] Socioeconomic factors were found to play a role, and male respondents were generally more sympathetic to the animal, but unlike the big cats worshipped by the Warli and other groups, the canids clearly don't have a significant fan base that could support conservation efforts.

Compared to leopards and tigers, snakes present a less obvious but ultimately deadlier threat to human survival. Causing an astonishing 45,900 deaths per year in India alone, snakebites count both as a globally neglected tropical disease and as a leading cause of injury-related deaths.

Félix Landry Yuan from the University of Hong Kong, China, and colleagues studied the attitudes toward snakes among the visitors in 30 sacred groves in the Western Ghats, a mountain range and biodiversity hotspot in the southwest of India.[26] Many but not all these groves are linked to the worship of snake deities.

The authors found that people were less likely to harm snakes when they encountered them within the sacred groves compared to encounters outside in their normal lives. Among those who would not harm snakes even outside the groves, the authors found many who worship snake deities, suggesting that, as with the case of Waghoba, the connection between the wild animal and the deity of their belief system protected the animal to a certain extent from being harmed in encounters with humans.

Landry Yuan and colleagues noted that the iconography of snake worship, often taking the shape of a king cobra (*Ophiophagus hannah*), goes back beyond the arrival of the current Indo-Aryan culture in South Asia, with the earliest evidence dating to around 2000 BCE. The researchers argued that venomous snakes are likely to have been a constant threat throughout human evolution, which explains the observation that deeply rooted cultural and religious beliefs evolved around their deadly power. Snake worship was widespread in the ancient world, but the Western Ghats are among the few places where it survived to the present day.

Another prominent example is the group of myths around the Rainbow Serpent found among many groups of Aboriginal peoples in Australia. The Rainbow Serpent is typically associated with creation stories and often also with fertility.

Again, the culture linked to snakes can be useful for conservation as well as for human safety. Snakes, including the king cobra (listed as vulnerable), are threatened by illegal hunting and trading. They are therefore protected under CITES (Convention on International Trade in Endangered Species of Wild Fauna and Flora) rules. Their special significance in ancient beliefs may inspire people to protect them from harm. Conversely, knowing about the prevalent snake species, their behavior, and dangers is an important step toward protecting the human population from the very real danger posed by these snakes.

Mesoamerica's pre-Columbian cultures lived closely with the jaguar (*Panthera onca*), the largest surviving feline in the Americas and a formidable hunter. All major civilizations, including the Olmec and the Maya, featured jaguar deities in their belief systems. Several cultures share the belief that the shamans derive their power from being able to transform into jaguars.

This may explain why there is widespread support for concerted efforts to preserve a contiguous jaguar habitat, the jaguar corridor, across its range, from Mexico to Brazil, led by the international feline conservation charity Panthera. The organization is active in 11 of the 18 countries of the jaguar's range: Belize, Brazil, Bolivia, Colombia, Costa Rica, Guatemala, Honduras, Mexico, Nicaragua, Panama, and Suriname. While the species is only listed as near threatened now, continued loss of forest habitat and prey species is darkening its outlook.

The close connection that Indigenous communities have with the jaguar will be an important part of conservation success. This aligns with the positive role these populations play in the stewardship of the surviving rainforest (see chapter 4).

While some species benefit from the combination of official, top-down conservation projects and deep-rooted cultural beliefs among the people at the front line, for some the cultural connection is the only hope. Such is the case for Sclater's monkey (*Cerco-pithecus sclateri*), one of only two primate species endemic to Nigeria. The monkey's very limited habitat in the Igbo-speaking south of the country does not overlap with any officially protected areas such as national parks. However, it does benefit from a ta-boo against killing monkeys still prevalent among local people, and from the habitat available in sacred groves in the area, some of which are specifically linked with the worship of monkey deities.

As Nigeria's population is rapidly growing and becoming more urbanized, habitat protected only by tradition may be at risk of be-ing lost to development. Moreover, the protection afforded by na-tive religions may diminish as Christianity gains ground in the area. Therefore, Lynne Baker from the American University of Ni-geria in Yola and colleagues surveyed the status of the monkeys and the sacred groves offering them shelter.[27]

To their surprise, the authors found that at the time of their sur-vey, in 2016, the groves they had previously surveyed in 2005 had only lost a moderate amount of tree cover, and none of them was cleared entirely. The monkey population in one specific site had even grown since the previous census.

In a postscript to the paper, however, Baker reported that after completion of the survey work, she learned that a significant part of one of the sacred groves had since been destroyed and converted to farmland without knowledge of community leaders. This expe-rience highlights the limitations of informal cultural protection, as it does depend on the benevolence of the local population and is not enforced by government.

Even when there is no religious element to the interspecies rela-tions, there may be a social one dictated by the plain practicalities of getting on. Thus, in the US state of Montana, where brown bears and grizzly bears are making a comeback, successful coexistence with the bears requires a degree of behavioral adaptation on the part of the humans. Measures minimizing the availability of easy food that might attract bears to human habitations include secure storage of food and waste. As Holly Nesbitt and colleagues at the University of Montana in Missoula pointed out in a study of human-bear coexistence in Montana, the effort to avoid any bear attractants is a collective action problem, only successful if everybody complies.[28]

Conversely, bears have also been shown to adjust their routines to the proximity of humans, some even switching to a nocturnal lifestyle to minimize conflict, as Clayton Lamb from the University of Alberta in Edmonton, Canada, and colleagues reported.[29]

Optimizing the peaceful coexistence of people with animals is important where wildlife is in retreat and conservation issues arise. On the other hand, a culturally ingrained habit of coexistence, whether it is based on ancient traditions or on improved understanding of the animals and their habits, is also important where species are recovering thanks to conservation efforts or where they have been re-introduced after prolonged absence in the course of rewilding projects. In Europe and parts of North America, for instance, where people have lost the traditional awareness of living alongside wildlife, it may benefit both sides to re-learn a culture of coexistence.

CHAPTER 8

Listen to Nature

Life on Earth remained silent for something like three billion years. Only a few hundred million years ago did evolution start to equip animals with anatomical features that produce sound, and with ears to hear them. Insects were once thought to have been the first sound makers, but new research has pushed back the origin of sound to the time when tetrapods conquered land. Dinosaurs marked a particularly noisy time, of which only the bird songs remain.

Today we see it as quite natural that many living things use sound and unnatural when nature goes quiet, as Rachel Carson anticipated in the title of her influential book *Silent Spring* (1962). Now, as populations of birds and insects decline, many people are noticing that once-familiar sounds of nature are getting quieter.

Thanks to recording and measuring devices for sound, ecologists can study the natural soundscape in quantitative and detailed ways, even without necessarily knowing which species are producing the sounds. This has given rise to the new research discipline of soundscape ecology.

Conversely, noises produced by our technology, from cars to ships, can also disrupt the natural soundscape and thereby affect animals, especially when they rely on sound for communication and orientation, as whales do. Thus it is worth paying attention both to the sounds that nature produces and to those that we produce and nature receives.

How life got loud

In the beginning, life was simple and silent. Single cells can't make sounds, and when multicellular plants and animals arose, at first only in the oceans, they quietly got on with the business of eating and trying not to be eaten.

The big question is, When did the hills come alive with the sound of nature's music? Science has long focused on the inescapable musicians of the animal kingdom and noted that birds, with their syrinx at the other end of the windpipe and completely unrelated to our larynx, appear to have started their singing careers independently from other branches. In that view, insects like today's crickets may have launched the terrestrial soundscape some 250 million years ago, followed by amphibians around 230 million years ago. Separate origins in different branches of the tree of life would seem to reflect the multiple origins of very different hearing organs. In between the noisy creatures, many other animal species appeared to have remained silent. But maybe we weren't listening the right way?

Results reported in 2022 suggest that more than 100 species from various clades previously assumed to be mute do in fact communicate acoustically, to the extent that the entire evolutionary history of vertebrate sound production may have to be re-assessed. Although reports of acoustic communications in unexpected species had been coming in sporadically, Gabriel Jorgewich-Cohen from the University of Zurich, Switzerland, and colleagues managed to move the dial by reporting in one go 53 additional species using sounds, including 50 turtle species representing more than half of all genera, as well as the evolutionarily unique tuatara (*Sphenodon punctatus*), one species of caecilian (*Typhlonectes compressicauda*), and the South American lungfish (*Lepidosiren paradoxa*). Adding a literature review of unexpectedly vocal species recently reported by others, the researchers arrived at a list of more than 100 species not previously known for communicating acoustically.[1]

Jorgewich-Cohen reportedly started sound recordings with his own pet turtles and expanded the endeavor from there, finding sounds wherever he checked. Feeding his new observations into phylogenetic tree models, he concluded that the ability to communicate with sounds is a homologous feature in all animals that have lungs, from lung fishes to all tetrapods.

While there are differences in the mechanisms used to start sound waves, most conspicuously in the birds' syrinx, Jorgewich-Cohen and colleagues note that all these animals use their lungs to power sound production—which sets them apart from noisy insects like cicadas. The authors further speculated that if vertebrate lungs were proven to share a common evolutionary origin with swim bladders in fishes, which in certain species also serve to power acoustic communication, the shared history of vertebrate sound production could be pushed back even further.

Whereas previous investigations, including a study from Zhuo Chen and John Wiens at the University of Arizona in Tucson, interpreted tetrapod sound production as a set of multiple-origin, convergent phenomena arising between 200 and 100 million years ago,[2] the newer analysis concluded that it likely evolved in our common ancestor with lung fishes, before tetrapods even conquered land. This would have been around 407 million years ago. Even adding just 12 turtles, the tuatara, and the lungfish to previous models would be sufficient to shift the result in favor of this early origin, Jorgewich-Cohen and colleagues argued. And if a link to the fish species producing sound with their swim bladder could be established, the shared vocal history could date back as far as 550 million years ago.

The key mistake that many made previously, the researchers noted, was to take absence of evidence as evidence of absence. They concluded that "the interpretation of acoustic behaviour as a non-homologous trait proposed in previous research was driven largely by a lack of information on key groups of animals."[3]

The technical differences such as the bird syrinx and the ultrasound apparatus of bats and dolphins may also have distracted from the issue. The authors argued, however, that all choanate vertebrates studied (vertebrates that have lungs and choanae, which are the passages that allow us to breathe through our noses) can make laryngeal sounds, including birds when they are hissing rather than singing. This insight creates a lot of room for further research into the functionality and ecological significance of the animal sounds we hadn't been aware of so far.

One use of acoustic communication in turtles that has been investigated in depth is the vocalization of hatchlings within their underground nest cavity. It has been hypothesized that the infant turtles use acoustic communications to co-ordinate their hatching, emergence, and run to the sea, giving each individual a better chance to escape any predators waiting for them.

Claudia Lacroix from the University of Toronto, Canada, and colleagues studied this vocalization behavior in the snapping turtle (*Chelydra serpentina*), a large freshwater turtle found in large parts of North America.[4] The researchers identified six different types of vocalizations, with one recorded even before egg pipping, and all six used in the 24 hours after. By manipulating the position of eggs in simulated nest environments, they could also show that it is an advantage for the development of hatchlings to be close to siblings. While this suggests that there is a benefit from group hatching as per the hypothesis, the authors were unable to trigger earlier hatching by playing recorded pre-hatching sounds.

Camila Ferrara from the Wildlife Conservation Society in Manaus, Brazil, and Jorgewich-Cohen made similar observations with the Arrau turtle (*Podocnemis expansa*)—one of the 50 turtle species in the study discussed above. The researchers recorded 11 different types of vocalizations and hypothesized that the baby turtles were communicating even before hatching.

Grown-up turtles may also be more vocal than was previously appreciated. Isabelle Charrier from the Institut des Neurosciences Paris-Saclay, France, and colleagues tagged free-swimming juvenile green sea turtles (*Chelonia mydas*) with microphones and recorded 10 different types of vocalizations, which they divided into four distinct categories: pulses, low-amplitude calls, frequency-modulated sounds, and squeaks.[5] These sounds were well matched to the previously determined hearing abilities of this species, so may well serve communication with conspecifics, although their functional significance remains to be established. This hypothesis is also supported by the finding that types of squeaks were shaped in ways specific to each individual, suggesting that turtles may recognize each other by their squeaks.

The authors call for further research on other marine turtles, which may reveal acoustic signaling underlying complex cooperative behaviors, such as the synchronized mass nesting of thousands of females on the same beach (arribadas) and the long migrations. Detecting and understanding their acoustic signals could also help with turtle conservation, for instance in helping to avoid bycatch, the authors pointed out.

Other new voices recognized in the growing choir of vertebrates include the crocodile newt (*Tylototriton himalayanus*), a species of salamander (Caudata) found in eastern India and Nepal. In contrast to anuran amphibians such as frogs and toads, salamanders in general were often considered to be mute, even though sporadic mentions of sounds are found in the literature. Documentary film makers and twin brothers Ajay and Vijay Bedi from India discovered during editing of material for their film *The Secret Life of Frogs* that they had recorded sounds produced by *T. himalayanus* during mating. With Robin Suyesh from the University of Delhi, India, they published an analysis of the reproductive behavior of *T. himalayanus* in the wild and characterized its vocalizations.[6] The authors described the sounds as short low-frequency calls, lasting around 70 milliseconds with a typical frequency of 656 hertz. Although salamanders lack external ears, it has been confirmed that they are able to sense low-frequency sounds. Specific experimental verification of the functional significance of these calls remains to be established.

Elsewhere in Southeast Asia, the Sunda gharial (*Tomistoma schlegelii*), a rare freshwater crocodilian categorized as vulnerable on the IUCN Red List, was always assumed to be a quiet reptile. Agata Staniewicz from the University of Bristol, United Kingdom, and colleagues studied its courtship and mating behavior using underwater sound recording equipment as well as video monitoring on animals held in captivity in Indonesia and in the United Kingdom and found a range of vocalizations.[7] The researchers documented seven different acoustic signals during courtship and mating, but none were detected outside this context. While the Sunda gharials don't engage in loud roaring like crocodiles, the most common sounds the researchers detected were low-frequency coughs lasting just 0.3 seconds, like the sounds described for American alligators (*Alligator mississippiensis*). Recordings also revealed sounds described as drums and rumbles, as well as moans, which only occur within call sequences combined with other sounds.

Within the often-noisy order of anuran amphibians, the genus *Crossodactylodes* (bromeliad frogs), represented by five species occurring in the Atlantic Forest in eastern Brazil, has long been considered mute. Marcus Thadeu Santos from the Federal University of Minas Gerais in Belo Horizonte, Brazil, and colleagues described the complex vocalizations of three of these species.[8] The researchers distinguished three call types they identified as creaking, chirp, and squeak calls.

As the Bedi twins noted almost apologetically in their paper, understanding reproduction is important for conservation. Therefore, it is useful if reproduction rites can be monitored with a simple microphone. The young but rapidly growing discipline of soundscape ecology has demonstrated that monitoring sounds can provide convenient access to information about the activities of multiple species where visual observation may not be possible, as we will explore later in this chapter.

The realization that a much wider range of vertebrates than previously thought are likely to be able to use acoustic communication also widens the scope for what can be studied with acoustic monitoring. All those species that were assumed to be mute may still have interesting stories to tell. Additionally, some species are discovered to communicate acoustically in a context not yet studied. Thus, Isabelle Charrier together with Andréa Thiebault from Nelson Mandela University in Port Elizabeth, South Africa, and colleagues reported underwater acoustic signaling used by penguins.[9]

At the same time, a wider-than-anticipated use of sound in nature implies additional risks for anthropogenic sounds to disrupt animal communication. Since the invention of steam engines, *Homo sapiens* has been the noisiest vertebrate species of all, blanketing entire landscapes with sounds and reaching volumes not attainable by natural organs, even producing supersonic booms by breaking through the sound barrier with airplanes.

Scientists have alerted the public to the problems of noise pollution in the context of whales and the disruption of their sonar by shipping noises, which can lead to disorientation and strandings. With many more species using sound, there are more communications that can be disturbed. We shall return to these problems at the end of this chapter.

Tuning in to bird behavior

Around 1715, the London-based printer and instrument-maker John Walsh published a book called *The Bird Fancyer's Delight, or Choice Observations and Directions concerning the Teaching of All Sorts of Singing Birds after the Flagelet and Flute [Recorder] When Rightly Made as to Size and Tone, with Lessons Properly Compos'd within the Compass and Faculty of Each Bird, Viz. for the Canary-Bird, Linnet, Bull-Finch, Wood-Lark, Black-Bird, Throustill [Thrush], Nightingale and Starling. The Whole Fairly Engraven and Carefully Corrected.* The short tunes, many of which are still reprinted in recorder collections today, reflected and catered to a pastime that is believed to have been popular at least since the Middle Ages—namely, to swap tunes with caged songbirds.

Many human composers have used or imitated birdsong in one way or another, from Jehan Vaillant's *Par maintes foys* in the fourteenth century through to contemporary composers like Emily Doolittle, who has written multiple works inspired by her interest in zoomusicology.[10] Conversely, several bird species, such as the European starling as well as the North American mockingbird, are often heard copying melodies produced by other species, including humans. Since the advent of sound recording, musicians have been able to incorporate samples of birdsong in their music as well as record duets with birds, as pioneered by cellist Beatrice Harrison (1892–1965). Harrison conversed with a nightingale in the BBC's first outdoor wildlife broadcast, coming live from a garden in Surrey, United Kingdom, in 1924. The resulting duet was so popular that the event was repeated every spring for the next 12 years. In April 2022, however, it was revealed that the BBC had secretly engaged a human nightingale imitator as a backup in case the bird didn't perform, resulting in some uncertainty as to whether the bird or the backup was heard in the pioneering broadcast.

One can debate which animal vocalizations may be called music and which should just count as calls, but regardless of musical merits, birdsong certainly has its unique value as a behavioral output signal that researchers can readily record and analyze. Multiple variables, such as the pitch, rhythm, and volume of the sounds, provide unique windows into the minds of birds as they respond to mates, rivals, or human disturbance.

Among the most impressive musical feats of songbirds is duet singing, which is observed in more than 200 species, including the plain-tailed wren (*Pheugopedius euophrys*), which is found in the Andes at altitudes between 2,200 and 2,300 meters (7,200 and 7,500 ft.). Males and females can each sing on their own, but they are often heard in antiphonal duets, meaning that they sing in turns alternating very rapidly, up to five times per second. Their co-operation is typically so well co-ordinated that any listener too far away to perceive the stereo effect will get the impression of just one bird singing. Choruses of more than two birds have also been reported.

Melissa Coleman from Scripps College, Pitzer College, and Claremont McKenna College in Claremont, California, and Eric Fortune from the New Jersey Institute of Technology in Newark, along with other colleagues, have studied this remarkable acoustic display since 2009. In 2021, they reported neurophysiological measurements conducted in the field with four pairs of duet-singing wrens.[11]

Specifically, the authors captured the wrens at a field site in Ecuador, implanted wire electrodes into their brains, and simultaneously recorded neurophysiological activity in the female and male of each pair while they sang. The brain region of interest is the HVC (now used as a common name, but originally an initialism for *hyperstriatum ventrale, pars caudalis*, which turned out to be an incorrect description). This is the area that has been identified as the center of song production in songbirds.

The researchers could show that HVC activity in a given individual increased when they were singing but decreased when it was their duet partner's turn. They conceptualize this regulation mechanism such that the heterogenous feedback, hearing the partner's sounds, inhibits the motor neurons responsible for singing, while autogenous feedback, the sound of one's own voice, doesn't.

As wrens of both sexes are known to give solo performances, the authors also measured how the HVC activity responds if a soloist is confronted with the voice of the partner in either solo or duet mode. Interestingly, the activity is inhibited in the female upon hearing male duet or solo syllables, but in the male only when he hears the female duet syllables.

In previous work, Susanne Hoffmann from the Max Planck Institute for Ornithology (now the Max Planck Institute for Biological Intelligence) in Seewiesen, Germany, and colleagues had detected alternating levels of HVC activity during duet singing in white-browed sparrow weavers (*Plocepasser mahali*), native to eastern and southern Africa.[12]

"[Hoffmann and colleagues] show the alternation in motor activity, as we show," Coleman explained. "We show that the regulation of the timing between the two birds is mediated by inhibition when hearing the partner. This sensory feedback effectively links the brains of the two individuals, which may be important for the coordination of the duet."[13]

Coleman and Fortune chose the duet singing not so much for musical reasons but as an accessible model for the neural regulation of co-operative behavior. "Our interest is the mechanics of co-operative performances, like dancing or singing together. We want to know the specific neural strategies that animal brains use to coordinate an individual's behavior with other individuals," Fortune explained. "The main feature of plain-tailed wrens that is useful for our analyses is that the birds rapidly take turns, allowing us to independently measure how the animals respond to cues from their partners and the signals that they generate themselves."[14]

The researchers suggested they were hoping to look for this feedback mechanism in other songbirds to see if it can be generalized. Findings may also impact the wider field of studying turn-taking in other animals.

If, as Eliot Brenowitz from the University of Washington in Seattle suggested, "song in both sexes may have been an ancestral trait in songbirds, with female song secondarily lost in some lineages as birds radiated from the tropics to higher latitudes," the impressive ensemble singing may be as relevant to the evolution of birdsong as the more widely heard solo performances.[15]

While duet singers are relatively rare and limited to the Southern Hemisphere, there are over 5,000 songbird species, typically featuring a male soloist showing off his musical skills to prospective partners and rivals, while females are silent in roughly half the species and show variable degrees of musical proficiency in the other half.

The production and learning of birdsong have already been studied for decades, but new insights can still be learned, and sometimes these studies even demonstrate connections to the music made by humans. The northern mockingbird, *Mimus polyglottos*, for instance, is widely known for its complex songs incorporating elements borrowed from other species or environmental sounds. As in many other songbird species, it is the male that sings to impress the female, so sexual selection is the most likely explanation for this complex display.

In sticking together borrowed sounds to produce a new and "original" melody, the mockingbird appears to work like a human musician, and one is tempted to think that the effect on the female is based on a sense of intrigue similar to ours when we're hearing sounds combined in a new way. However, behavioral studies of how complex songs like the mockingbird's appeal to the female audience are still lacking. So far, only simple elements have been tested, such as "sexy syllables" in canaries. However, an attraction to specific sounds cannot explain the sexual selection favoring complex and original compositions.

Neuroscientist Tina Roeske from the Max Planck Institute for Empirical Aesthetics in Frankfurt, Germany, musicologist David Rothenberg from the New Jersey Institute of Technology, and ornithologist David Gammon at Elon University, North Carolina, have analyzed sound recordings of mockingbirds to identify elements of complexity that might be stirring the interest of the female audience.[16]

Comparing elements in the output of specific mockingbirds, Roeske and colleagues found that the birds, like human composers and musicians, tend to repeat short sequences and then change them in a limited way. At least in the human listener, this habit grabs the attention. We recognize a pattern, we think we know where it's going, but just when we are sure we've understood it, the composer introduces a variation that takes the music to a different place.

The mockingbird won't turn a borrowed idea into a fugue or a sonata, but it does change it in ways that have equivalents in human music. Roeske and colleagues identified four separate ways in which mockingbirds change the original syllable or phrase, which they describe as "morphing." In repeating a phrase, the birds may change the timbre or the pitch of the sound; they may slow it down or speed it up. For each of these effects, the authors identified equivalents in human music, which they juxtaposed in a video explaining their findings.[17]

As Rothenberg explains in the video, the perception that the mockingbird just mimics other birds is an oversimplification. "The mockingbird composes very carefully his own song out of the songs of other birds," Rothenberg says. The analysis demonstrated some of the rules the birds follow in this compositional process. Conceivably, experimenters could now test these compositional strategies separately to investigate their effects on the females. The authors admitted that one would also have to consider the possibility that some of the morphing could be due to simple physical constraints of song production as opposed to an evolved strategy to make the song interesting.

Birdsong has also been compared to human languages. One organizational principle that both modes of communication share is Menzerath's law, which predicts that elements with more constituents tend to consist of shorter constituents. In 2021 the group of Jon Sakata at McGill University in Montreal, Canada, reported that across 15 species of songbirds, complexity patterns following Menzerath's law were observed.[18] Remarkably, the pattern is not disturbed in birds that were not taught by their parents. This, the authors concluded, may mean that physical constraints, such as motor production biases, contribute to this pattern.

The zebra finch (*Taeniopygia guttata*) is the most widely used model species for studies of birdsong and its biological foundations. In one such study, Coen Elemans's group at the University of Southern Denmark, in Odense, Denmark, showed how the innervation of single muscle fibers and low muscle stress enable the precise control of pitch to a resolution of less than one hertz.[19] This remarkable level of control expands the realm of possibilities of acoustic expression and may thus have contributed to the rapid radiation of songbird species, which now represent nearly half of the current avian biodiversity.

Among zebra finches, only the male produces song, but the female can also remember her father's song, as behavioral studies have shown. Using the related Bengalese finch (*Lonchura striata* var. *domestica*), a widely traded cage bird domesticated in Asia, Kazuo Okanoya's group at the University of Tokyo, Japan, set out to investigate the finch's memory of its father's song in both sexes and throughout the bird's lifespan.[20] The authors found that females maintained the same level of responses, indicating recognition of their father's song throughout their lives. Although they never sing, it is thought that remembering songs has a selective benefit in that it helps female choosing a mate based on song quality.

In the males, by contrast, their propensity to sing while a recording of a song is played to them can serve as a measure of their response. While they are young, they fall silent to pay close attention to their father's song, presumably to learn from it, but in adulthood they are more likely to ignore the other song and continue singing their own, as if to take a stand against a male competitor.

Beyond the level of individual behavior, birdsong enables researchers to monitor activity at the population level. In the emerging field of soundscape ecology, birdsong is among the natural sounds that reflect the response of wildlife to natural events, such as hurricanes, as well as to human disturbances. It has the added benefit of being coupled with the circadian clock. Like the famous examples of the nightingale and the lark, each bird species has its characteristic time profile. Thus, when disturbances like traffic noise or light pollution disrupt the birds' circadian cycles, this can be detected and quantified with sound recordings.

Conversely, when we hear the birds singing as they should, this contributes to our experience of nature. Clinton Francis's group at California Polytechnic State University in San Luis Obispo demonstrated the effect by exposing hikers to "phantom birdsong chorus" with the recorded voices of a diverse group of birds.[21] The researchers used two separate but similar hiking trails where the sounds could be switched on and off to control for differences of time and place. They found that hikers who had been exposed to the extra dose of birdsong reported a higher degree of a restorative effect and general well-being. Maybe our affinity for music does have its deepest roots in the singing of the birds.

Cultured cetaceans

Fans of traditional music can walk into an Irish pub in Oxford or Melbourne, in Germany or California, or indeed anywhere else in the world where music isn't banned by authorities, and if they find people playing music there, they will likely recognize some of the tunes. If they remember to bring their instrument, they can even join in.

Traditional melodies have been passed down the generations for centuries, and with migration and globalization they have also spread around the world. Celtic music, in particular, which includes Irish, Scottish, Breton, and Galician traditions among others, has become a truly global phenomenon, which is celebrated annually at the Festival Interceltique in Lorient, Brittany.

In contrast to the way we learn the more academic forms of knowledge, from classical music through to mathematics, which are also spread globally but tend to rely on codified material written down, many folk musicians reject printed versions and prefer to learn from listening. Although some classically trained musicians are frightened by the mere thought of playing without a score, this way of learning isn't as difficult as it looks. There are two strategies. The first works much like singing along to a song that often plays on the radio. You essentially make it up as you go along, trying to get as many notes right as you can. There are more important notes that will be more obvious, like a catchy chorus, and you can start by concentrating on those and fit in the other ones later, until, maybe months later, you realize you can play the tune without ever having consciously memorized it.

The slightly more thorough and demanding, but also quicker, way to learn is to find a patient teacher who will cut the tune into pieces short enough to be remembered after a few plays, which are then repeated and finally stuck together. In the absence of living people to learn from, both methods can be achieved in solitary sessions with a recorded track or video found online, reconciling the traditional approach with modern technology.

These techniques are not only practiced on all continents—they are also used in the oceans. Research has shown that humpback whales (*Megaptera novaeangliae*) proceed similarly when they learn a new song. Humpback whales produce elaborate and extended songs that can last up to 30 minutes and are sometimes repeated for hours. Like in human music, the vocalizations are constructed with a complex pattern of repeated phrases organized in themes that then combine to constitute the song. Humpback populations spread out over large areas share the same song, which evolves like a cultural tradition. On rare occasions, the traditional song can be replaced rapidly by a revolutionary takeover of a new song picked up from a neighboring population.

Thus far, there are only a few known "song revolutions," for which five sound recordings of transition songs exist. To elucidate how the animals learn new tunes, Ellen Garland from the University of St Andrews, United Kingdom, and colleagues conducted a detailed analysis of four recordings of the hybridized songs that humpback whales produced during these transitions, while excluding a fifth due to insufficient sound quality.[22]

It is known that humpback whales are vocal production learners, meaning that they adjust their vocalizations in response to what they hear from other members of their population, much like folk musicians playing by ear. What Garland and colleagues sought to establish was whether the whales also cut down the new song into segments to learn these more rapidly.

The researchers found that whales do indeed segment the new song into manageable pieces, which they gradually incorporate into their traditional display. This appears to be facilitated by transitional phrases used in places where old and new tunes share some similarities. As they incorporate more parts of the new song, they progressively delete the previous tune, ensuring that the entire population conforms to a single tune.

Meanwhile, the functional significance of these songs remains to be elucidated. The fact that only males sing them appears to suggest that it may be to do with sexual selection. The immense effort that goes into keeping the uniformity suggests it is important. It is one of the more spectacular and widely known cultural phenomena among whales but by no means the only one. Humpback whales also display social learning of their migration routes and feeding strategies, although for these issues toothed whale species are more rewarding study subjects.

Listening to the sounds of the biosphere

Hurricane Irma passed around 100 kilometers (62 mi.) north of Puerto Rico on September 6, 2017. Just two weeks later, the island took a direct hit from the even more powerful Hurricane Maria.

Both storms ranked among the most powerful Atlantic hurricanes recorded up to that point and caused widespread devastation and numerous deaths along their paths through the Caribbean.

Just half a year before, Ben Gottesman from the Center for Global Soundscapes at Purdue University, Indiana, and colleagues had set up acoustic monitoring equipment in coastal forests and coral reefs of Puerto Rico's southwest coast. They were hoping to record the acoustic profile of the changing biodiversity from the forests to the sea, but the tropical storms provided them with a unique dataset showing how nature responds to a major disaster—or two.

The recordings showed a range of different effects in the aftermath of the storms. Snapping shrimps, which normally produce one of the loudest noises known in marine biology, were silenced during Hurricane Maria and took several days to resume their daily cycle with snapping peaks at dawn and dusk.

By contrast, the nightly fish chorus, which helps fish finding mates and defending spawning territories, but may also attract the attention of predators, became bolder during Hurricane Irma. Gottesman and colleagues have suggested that the increase in ocean water turbidity caused by the passing hurricane may have shielded the fish from predators and thus enabled them to be more vocal.

In the coastal dry forests after Hurricane Maria, the recordings show a marked reduction of insect noises that lasted for about three weeks. This work shows how simple sound recordings can monitor nature's response to environmental change. In this example of short-term impacts from catastrophic storms, both terrestrial and marine environments recovered their soundscapes within weeks.[23]

Chronic disturbances such as land-use change or climate change, however, may very well lead to permanent changes in the natural soundscape. One of the pioneers of soundscape ecology, Bernie Krause, a professional musician and sound engineer based in Glen Ellen, California, used sound monitoring over many years to demonstrate biodiversity loss, which may very well manifest as a "silent spring," as predicted by Rachel Carson.

Krause defined the essential vocabulary of the field, including *geophony, biophony,* and *anthropophony,* for sounds produced by inanimate, biological, and human-made processes, respectively, as well as the *acoustic niche,* representing the acoustic equivalent of an ecological niche.

The increasing exposure of marine environments, including the hitherto pristine polar waters, to ship traffic and the concomitant noises are widely recognized as a threat to marine mammals such as whales, which tend to rely on sound for navigation and communication. Therefore, eco-acoustic studies in the marine environment often focus on the acoustic behavior and needs of one species and the way in which it interacts with the wider soundscape, especially if the latter is changed by human activities. Aran Mooney from the Woods Hole Oceanographic Institution, Massachusetts, and colleagues, for instance, studied the hearing ability of a wild population of beluga whales (*Delphinapterus leucas*) and analyzed the soundscape of its natural environment.

The researchers used techniques derived from hearing tests used for infants and adapted them to the marine environment such that the sensors used to pick up brain activity could be attached to the whales' heads with suction cups.[24] Compared with previous studies on bottlenose dolphins and humans, they found that belugas appear to be less likely to suffer hearing loss with age. This may simply be due to both humans and dolphins living in noisier environments than beluga whales.

They then went on to characterize the acoustic environment in the population's habitat in the Nushagak Estuary in Bristol Bay, Alaska.[25] They found that their very sensitive hearing enables the belugas to pick up the full range of sounds present in their habitat, including very quiet sounds.

Narwhals (*Monodon monoceros*) also rely on sound for echolocation and feeding and have increasingly found attention as a sen-
tinel species for the impact of climate change in the Arctic. As
we've already seen in chapter 2, an acoustic analysis by Jens Koblitz from the Bioacoustics Network, Germany, and colleagues
found that the narwhal echolocation clicks are emitted as a very
strong and extremely focused beam.[26]

The researchers concluded that the narwhals are well adapted
to hunting with echolocation in the geologically unique "acoustic
bowl" of Baffin Bay, where most of these animals feed during win-
tertime. Echolocation enables them to navigate and find prey in
permanent darkness at great depths, diving beyond 1,500 meters
(4,900 ft.). When they come back up for air, the focused beam also
enables them to find gaps in the dense ice cover, which will have
moved since they started the dive.

With increasing shipping in the Arctic, however, there is concern that anthropogenic noise will affect these narwhals, especially in their summer feeding grounds in Lancaster Sound, which
is close to the Northwest Passage.

Susanna Blackwell from Greeneridge Sciences Inc. (a company
based at Santa Barbara, California, producing acoustic sensors)
and colleagues reported extended acoustic monitoring and satellite tracking of six narwhals in Scoresby Sound, East Greenland,
for up to seven days.[27] The researchers found that the narwhals
stayed silent for around a day after release, highlighting the importance of longer-term observations.

When the animals resumed their normal activities, the monitoring showed that they produced echolocation buzzes mainly at
depths of 350 to 650 meters (1,150–2,130 ft.) and especially in one
specific fjord, which may be their preferred feeding location. Social
calls, by contrast, were typically produced closer to the surface.

Anthropogenic noise impacts the soundscape experienced and used by whales by reducing the space across which the animals can communicate with sound. In a model study, Danielle Cholewiak from the NOAA (National Oceanic and Atmospheric Administration) Northeast Fisheries Science Center at Woods Hole, Massachusetts, and colleagues quantified this effect for four different species of baleen whales in a marine protected area, the Stellwagen Bank National Marine Sanctuary.[28] A previous analysis in 2012 had shown that some North Atlantic right whales (*Eubalaena glacialis*) may have lost two-thirds of their communication space when foraging in essential habitat, based on a comparison of current ambient noise with an estimate of pre-industrial noise levels.

Cholewiak and colleagues extended the analysis to be able to compare the impact of noise on five different communication modes of four baleen species, including fin whale song, humpback whale social sounds, humpback whale song, minke whale pulse trains, and right whale gunshot sounds, and also to differentiate between the contributions of different kinds of vessels, ranging from large commercial vessels, for which satellite tracking data is routinely available, to smaller, locally based vessels used for fishing or whale watching.

The researchers found that the larger vessels contributed most to the disturbance and that some of the species had their communication space reduced by 80% compared with pre-industrial noise levels.

In separate work, Rosalind Rolland at the New England Aquarium in Boston and colleagues have shown that right whales suffer symptoms of chronic stress linked to ship noises. This work was originally conceived as a general investigation of health and hormones levels in whales, but the shutdown of traffic after the attacks of September 11, 2001, provided the opportunity to compare the hormone levels in the absence as well as in the presence of shipping.

More recently, in a study of the foraging behavior and acoustic environment of humpback whales, Hannah Blair from Syracuse University, New York, and colleagues also found evidence suggesting that the presence of shipping noise leads to stress in the whales.[29]

New discoveries around the importance of sound for the lives of cetaceans are still being made quite regularly. For instance, while humpback whales (*Megaptera novaeangliae*) are known for their elaborate songs that are spread around the population by social learning, as we discussed earlier in this chapter, Kate Stafford from the University of Washington in Seattle and colleagues from the Norwegian Polar Institute in Tromsø reported in 2018 that bowhead whales off the coasts of Spitsbergen are just as musical but more inclined to innovation, creating a jazzlike freeform sound and new songs every year.[30]

The research showed that bowhead whales are a kind of musician you may not want to have as a neighbor—they sing loudly, 24 hours a day, from November through to April. As their singing season falls in the Arctic winter and mainly happens under ice cover, it had remained unknown until it was discovered with the help of new hydrophones that can be deployed under these conditions. The evolutionary and ecological importance of the bowheads' all-winter jazz session (i.e., their acoustic niche in the wider marine soundscape) remains to be uncovered.

One drawback of the vocalizations used by many marine mammals is that predators may also hear them and follow the sound to find their prey. This is a risk, for instance, for the young calves of the southern right whale (*Eubalaena australis*), which are targeted by eavesdropping killer whales (*Orcinus orca*).

Mia Nielsen from the University of Aarhus, Denmark, and colleagues sought out nursing right whale mothers and their calves in breeding grounds on the coast of Western Australia to investigate how they adapt to the threat of being heard by predators.[31]

Southern right whales migrate over large distances between feeding grounds at higher latitudes and breeding grounds with warmer water. With their young calves, they tend to stay in very shallow coastal water, where the breaking waves provide both visual and acoustic cover. Attaching microphone tags to whale mothers and monitoring their communications, the researchers found that whales relied on close contact and called as little and as quietly as possible, with just under one call recorded per dive. Similarly cryptic communications have been observed in breeding humpback whales.

Even with the efforts of snapping shrimps and bowhead whales, the oceans tend to be a relatively quiet place. In terrestrial ecosystems, however, sounds produced by many vociferous species tend to compete with each other as well as with the geophony of the wind and water cycle, and with the increasingly ubiquitous noise emitted by humans and their machines.

With such a complex mixture of sounds, recordings are easy to make and can be interpreted in terms of loudness and diversity, but deconvolution for detailed interpretation is likely to become more challenging.

In a proof-of-principle study of the feasibility of ecosystem monitoring via sound recording, Nick Friedman and colleagues from the Okinawa Institute of Science and Technology, Japan, set up a network of 24 long-term acoustic monitoring stations along a gradient of urban to rural environments on the island of Okinawa. This acoustic monitoring is part of a wider environmental monitoring project that also includes cameras, arthropod traps, and environmental and weather recording facilities.

A preliminary report from this project used data collected at five of these sites over one month to establish the possibilities and challenges to be encountered in the analysis of such datasets.[32] For this investigation, the stations recorded 10 minutes of audio every half hour. Using machine-learning approaches for automatic species recognition, the authors managed to implement reliable recognition of several bird species, including two that are of conservation interest and high cultural significance.

On the other hand, acoustic parameters don't always align with biological parameters in the expected ways. For instance, acoustic diversity as defined by the diversity of one-kilohertz frequency bands observed, did not correlate with biodiversity. Instead, this parameter was found to decrease with increasing forest cover.

Moreover, some of the observation parameters were prone to being swamped by highly abundant and noisy species such as cicadas. Therefore, the authors emphasized the need to cross-correlate the acoustic recordings with other kinds of information gained from the observation network to ensure the acoustic changes are interpreted correctly. With this, state-of-the-art machine-learning technology should also prove very valuable. In a follow-up paper in 2021, the authors made specific suggestions on study design and which acoustic indices to use for monitoring biodiversity.[33]

In terrestrial soundscapes, bird song is the most conspicuous and widely studied sound. Recordings can be readily analyzed to look for details of species diversity, population dynamics, circadian clock behavior, and much more, as well as to monitor these parameters as they change over time and are affected by human activities.

Birdsong is part of the wider terrestrial soundscape and interacts with the acoustic environment in multiple ways. Forests, for instance, can help or hinder the propagation of the sounds, depending on growth and grazing.

Megan Gall's group at Vassar College, New York, has shown how grazing by white-tailed deer (*Odocoileus virginianus*) can change the acoustic properties of woodland habitat in temperate climate zones.[34] Acoustic measurements showed that sound fidelity but not amplitude is improved by grazing. The finding suggests that songbird vocalizations might be more effective and transmit information more clearly in deer-browsed environments. This effect could benefit both intended receivers and unintended receivers such as predators. The real ecological impact of such changes in the acoustic environment remains to be established.

Birdsong can also help with research into the separation of species, as birds tend to respond differently to sounds they perceive as coming from a conspecific, compared to sounds of other bird species.

Jason Weir from the University of Toronto, Canada, and Trevor Price from the University of Chicago used birdsong recordings to study the speed of species separation in the Amazon as compared to temperate zones of North America. Using DNA comparisons to estimate how long pairs of populations have been separated, the authors calculated the time of separation needed for the aggression against the song of a former conspecific to decline to half its original extent. The results differed dramatically, with an average of three million years determined for 51 pairs of populations in Amazonia versus half a million years for 58 pairs of populations in North America. If this result reflects actual speciation rates, it implies that the high diversity of birds in the Amazon cannot be due to a high rate of speciation. Instead it must be assigned to a lower rate of species loss.

The steady seasonal change of the natural soundscapes has been monitored in Alaska, where a network of 62 monitoring stations have been set up at the Kenai National Wildlife Refuge.[35] Meanwhile, applying the approach to the study of more abrupt change, Amandine Gasc from the Center for Global Soundscapes and colleagues have studied the aftermath of a devastating wildfire that occurred at Chiricahua National Monument, Arizona, in 2011.[36]

Technology enabling the distributed recording of sound at many sites and over long periods is beginning to prove helpful as a way of monitoring environmental change. The Center for Global Soundscapes, for instance, has set up a citizen science project through the app Record the Earth (available for iOS and Android) with the aim of establishing a global sound archive.

Sophisticated machine-learning analyses of such data sets may even be able to pick up species decline before it becomes detectable with conventional methods. Thus, listening closely to the complex orchestra of species may help us prevent the catastrophe of the biosphere one day falling silent.

Human-made noise rocks the oceans

For decades, the technology to harvest polymetallic nodules (which contain nickel, cobalt, copper, and manganese) and massive sulfide systems (which are rich in copper and gold) from the seafloor has been promised as being just around the corner.

On June 30, 2021, it came considerably closer to becoming a reality, as the Pacific island state of Nauru triggered the two-year-rule in the 1994 agreement related to the United Nations Convention on the Law of the Sea (UNCLOS) to force the International Seabed Authority (ISA) to finalize regulations for seafloor mining by the end of June 2023. As the ISA failed to agree to a regulatory framework by then, Nauru could allow companies to proceed without one, bound only by existing international law. However, at the July 2023 ISA meeting the issue was deferred to 2024.

The seafloor of interest is the Clarion-Clipperton Zone, a vast area around the Clipperton Fracture stretching across the North Pacific between Hawaii and Mexico. While the area has an abundant supply of polymetallic nodules literally lying around on the seafloor, it is also home to an unexplored biodiversity.

Nauru had teamed up with the company DeepGreen, which was then acquired by the Metals Company (TMC). The company promises to deliver lower-impact battery metals for the clean-energy transition by using metal resources from polymetallic nodules, and ultimately it promises to stop mining when enough metals are available for a circular metal economy.

Critics have noted, however, that the economic history of Nauru doesn't exactly inspire confidence, as the island nation has devastated most of its land surface in the name of phosphorus mining, and when that resource ran out, it hosted offshore detention camps for refugees arriving in Australia. Kiribati and Tonga are also working with TMC to develop mining projects, while other Pacific island nations, such as Micronesia, Samoa, Fiji, and Palau, are opposing such plans.

Numerous conservation organizations and even companies like BMW and Google are supporting a moratorium on deep-sea mining on the grounds that the seafloor is so unexplored that we wouldn't even know what ecosystems and species will be destroyed by mining operations. When samples were taken in 2016, half the species identified were new to science. There were also indications that some species may attach themselves to the polymetallic nodules and use them in some way, meaning that their removal would impact the ecosystem.

Concerns over the disturbance of the actual seafloor communities have already been raised many times and widely discussed. In 2022, researchers pointed out a separate problem for marine ecosystems that had not been addressed yet—namely, noise pollution. Rob Williams from Oceans Initiative in Seattle and colleagues estimated noise emissions of the proposed mining activities in the Clarion-Clipperton Zone.[37] According to the authors, there is no publicly accessible data available for the noise emissions of mining equipment operating on the seafloor. Even the environmental impact statement of the Solwara project of the now defunct company Nautilus only includes modeled sound data for the surface vessel.

Due to the lack of data for seafloor mining equipment, the researchers used analogues such as coastal dredging equipment removing gravel and sediment. For each of the booster pumps that transport mined minerals up the riser to the surface, comparable pumps of coastal dredging operations were considered. The authors expected that the results would be underestimates, as the shallow-water surrogates would likely miss out on low-frequency noises produced by the heavier deep-sea equipment.

Even with this incomplete information, Williams and colleagues calculated that the mining and pumping operations will produce a vast cylinder of sound propagating outward. This noise will exceed ambient levels of sound typical of quiet weather conditions in a radius of 500 kilometers (310 mi.). Within a radius of 4–6 kilometers (2.5–3.7 mi.), the noise would exceed a threshold determined by the US National Marine Fisheries Service to cause behavioral impacts on marine mammals.

The authors pointed out that the long-distance reach of sound makes it impossible to determine the ecological impact of the mining by comparing the mined area to similar nearby regions not affected by the operation. To be outside the reach of the noise produced, unimpacted areas would have to be so far away that they would host completely different ecosystems and would therefore not be able to offer a useful comparison.

Given these fundamental difficulties, the authors called for more transparency from mining companies: "We urge contractors to release in a timely manner information on sound-source characteristics of all seabed-mining components." They also called for the ISA to set thresholds at a precautionary level. They concluded: "We recommend that mining should be coupled with intensive, independent studies of regional baselines and environmental impacts, as well as with the ability to alter or halt operations quickly if newly acquired data indicate notable unexpected effects."

Marine noise pollution is known to be a problem for cetacean species in particular, many of which use sound for communication and navigation, as discussed above. Deep-diving toothed whales are considered particularly vulnerable. Stranding events of these cetaceans have often been linked to acoustic disturbances, but mechanisms of causation and reasons for different sensitivities between species and populations remain to be found.

One hypothesis being investigated is that adaptations that evolved to avoid predators might drive the responses to anthropogenic noises. Specifically, the risk-disturbance hypothesis predicts that the response to an acoustic disturbance is guided by the trade-off between the perceived predation risk and the cost of missed foraging opportunities.

To test this hypothesis, Patrick Miller from the University of St Andrews, United Kingdom, and colleagues exposed four species of cetaceans to anthropogenic sounds as well as to recorded sounds of their predators, killer whales (*Orcinus orca*).[38]

The authors studied northern bottlenose (*Hyperoodon ampullatus*), humpback (*Megaptera novaeangliae*), sperm (*Physeter macrocephalus*), and long-finned pilot (*Globicephala melas*) whales in their feeding grounds and measured reductions in foraging time in response to the sound exposure. These four species use different strategies against killer whale predation. Adult sperm and humpback whales can rely on their fight capability, and pilot whales use group mobbing tactics, while bottlenose whales tend to hide from the predators even at the cost of having to abandon foraging.

Miller and colleagues found that across these four species, the behavioral response to killer whale sounds is a good predictor of the response to naval sonar. As expected, northern bottlenose whales, being most vulnerable to predation, had the strongest and most consistent avoidance response to sonar, even though humpback whales are the ones with the most acute hearing in the frequency range of the sonar used. Humpbacks only avoid the sounds when they have vulnerable calves with them. Sperm and long-finned pilot whales are least affected by the sounds. Results of an earlier study with Blainville's beaked whale (*Mesoplodon densirostris*) fit the correlation in that they stopped feeding in response to the sounds and are considered vulnerable to orca predation.[39]

The authors concluded that all cetaceans relying on predator avoidance are likely to have costly avoidance responses to anthropogenic noises too, including the harbor porpoise, narwhal, beluga, minke, and sei whales, as well as beaked whale species. Several of these species had already been shown to strongly respond to noise.

There is the possibility that cetaceans will eventually learn to distinguish between the sounds produced by killer whales and humans. However, the authors warn that the increasing advances of both species into the Arctic as the sea-ice cover shrinks will hit naïve populations hard: "We extrapolate that these Arctic specialists will face a looming double-whammy impact of increased direct predation and potentially severe maladaptive responses to novel anthropogenic sounds."

One of the species most at risk from this double whammy is the narwhal (*Monodon monoceros*), whose use of sound we discussed earlier in this chapter. For more specific and mechanistic insights into how noise affects individual cetaceans, Terrie Williams from the University of California, Santa Cruz, and colleagues equipped narwhals with custom-built monitors recording heart rate, acceleration, and depth, and they observed the whales' physiological response to anthropogenic noises.[40]

Two of the 13 narwhals studied were exposed to pulsed sounds of two air guns used for a seismic measurement program. The guns were towed behind a ship, floating 6 meters (20 ft.) below the surface, and fired synchronously every 80 seconds as the ship moved into and out of fjords to be measured. The two animals exposed showed an intense and somewhat paradoxical stress response. They glided 80% less during diving descents, swimming with more than 40 strokes per minute, while their heart rates dropped below 10 beats per minute; their breathing at the surface was 1.5 times faster. The low heart rate (bradycardia) is a typical fear response—even lower rates were measured in narwhals entangled in nets.

Overall, this physiological response, combined with behavioral responses such as decreased foraging and fleeing from open water to coastal areas, is very costly in terms of energy consumption, the researchers pointed out. They calculated that a startle dive in response to seismic noise required twice as much energy as a matched control dive.

As the disappearance of sea-ice cover is enabling more shipping and resource extraction activities in the Arctic, narwhals will face more of this acoustic stress in the near future, and it may affect their ability to survive.

From the South Pacific to the Arctic, the marine soundscape is changing everywhere due to the progressing industrialization of the oceans, the decline of sound-producing species, and physical changes such as the loss of sea ice, as Carlos Duarte from the King Abdullah University of Science and Technology in Thuwal, Saudi Arabia, and colleagues pointed out in a review of the "soundscape of the Anthropocene ocean."[41]

The authors found that, beyond the known effects on marine mammals, anthropogenic noise impacts fish and invertebrate species and even seabirds and reptiles. An aggravating factor arises from ocean acidification. The increasing CO_2 concentration in sea water enables sound to travel even farther than it could previously.

On the positive side, however, noise is uniquely different from other human disturbances in that we can switch off the source and the impact will cease immediately.

Sophie Nedelec from the University of Exeter, United Kingdom, and colleagues, including review co-author Stephen Simpson, studied recovery after noise removal for the case of coral reef ecosystems affected by motorboat noise.[42] These authors studied reproductive success of the fish species spiny chromis (*Acanthochromis polyacanthus*) on the reef during a breeding season when motorboat traffic was restricted and found that the population produced more and bigger juveniles. The researchers conducted additional laboratory breeding experiments to isolate sound as the causative factor.

Similar recovery effects were observed when human activities ceased during COVID-19 lockdowns. Duarte and colleagues called for recognition of noise pollution as an important anthropogenic factor impacting marine ecosystems. They proposed a range of measures that could mitigate the problem. For instance, retrofitting large container ships to reduce propeller cavitation (formation of small vapor-filled cavities in the water) has been shown to reduce both fuel costs and sound emission. A switch from fossil fuels to electric motors in shipping would also reduce the noise produced.

However, these improvements may be dwarfed by increases of noise pollution if seafloor mining plans go ahead and further industrial activities expand into the melting Arctic. The conditions established for mining will be crucial, and there is also a case to be made for reconsidering the necessity of seafloor mining. The professed requirement of the metals for batteries in electric vehicles may soon become obsolete as new battery technology becomes available. As Andrew Thaler of Blackbeard Biologic explained in the DSM Observer, "Deep-sea mining is in an ever-tightening race against next-generation solid-state batteries."[43] These will be favored by the industry because they achieve higher charge density and shorter charging times. A collateral benefit is that they don't use any exotic metals for which one would have to start mining the seafloor.

Regulating industrial activity on the seafloor will be necessary and may reduce harm, but the best outcome for the marine environment could be achieved if we just limited our seafloor activities to scientific exploration.

CHAPTER 9

Animals Shaping the Environment

We humans have shaped and reconfigured our planet in dramatic ways, such as connecting oceans with canals, removing mountains, clearing forests, and creating reservoir lakes. While these measures were intentional (if sometimes reckless), we are now realizing that we are producing a lot of unintended changes as well—such as melting the polar ice caps and raising the sea level with our emissions of carbon dioxide from burning fossil fuels.

In our enthusiasm for modern technology and infrastructure, we have often paid too little attention to what was there before. Colonial powers, in the drive to bring their version of civilization to other continents, destroyed other civilizations. Similarly, in efforts to shape our environment to better extract its resources, we are destroying natural ecosystems and complex webs of material transport and recycling. Some of these are so subtle and intricate that we overlooked them for centuries and only came to notice them in the last few decades, midway into their destruction.

For instance, whales, many species of which narrowly escaped extinction by industrial whaling (as we saw in chapter 5), as well as seabirds and salmon are all part of a global nutrient recycling network that works against the direction dictated by gravity and thus saves vital nutrients from disappearing in the seafloor sediment. Similarly, vultures work in the recycling business and thereby keep us safe from certain diseases, while earthworms keep the soil, which we need to grow our food, fertile.

All these elements, as we will see in this chapter, are part of ecosystem services that we are already destroying. As with the hunting discussed in chapter 5, we have been busy sawing off the branch we're sitting on—partly because we didn't sufficiently understand and appreciate the ecological networks of life on Earth.

Megafauna moves nutrients uphill

Around the middle of the nineteenth century, the Chincha Islands, some 21 kilometers (13 mi.) off the Peruvian coast near the town of Pisco, were the most valuable piece of land in the world. This was not due to their natural beauty and the spectacular views of the Andes and the Pacific. The land was covered with a layer, several meters in height, of bird droppings, also known as guano. Chincha Islands guano was considered the best in the world, and it launched a rush that led to every bit of it being scraped off the rocks within a few decades. Working conditions for the enslaved people harvesting the guano were dismal, corruption was rife, and the Peruvian authorities squandered the wealth gained in the guano rush so fast that the country went bankrupt when the guano ran out.

There were other locations where guano accumulated, but in most of them there was too much rain washing out the soluble nitrogen compounds, leaving a less than perfect fertilizer. The rapidly growing world population was once more facing the threat of mass starvation—until Fritz Haber and Carl Bosch developed industrial synthesis of ammonia from the elements, which was first used at industrial scale in 1913 to produce fertilizer out of nitrogen taken from the air. Today this process converts as much nitrogen as natural fixation, meaning that it supplies roughly half the nitrogen that keeps humanity alive.

With the industrial production of fertilizer, humans dramatically changed the distribution of nutrients in the biosphere—before we even began to understand how this distribution worked before our species conquered the world. Only now are scientists beginning to realize that the seabirds leaving their droppings on the shores may just be one link in a chain of nutrient distribution reaching from the deep sea all the way up to the high mountain slopes. This system relies on large animals, many of which are either extinct already or under threat.

Given the laws of gravity and the hydrological cycle, there is a strong likelihood that nutrients available on land, even though they may go through many cycles around the food web, will eventually be washed out to the sea. In the oceans, there is the risk that they will drop out of the photic zone and reach the seafloor, where they will be buried in sediment that may only be returned to circulation on geological timescales, some tens or hundreds of millions of years later.

Animals can make important contributions to stem this flow, as was first reported for whales back in 2010. Researchers studying the ecology of sperm whales found that they harvest nutrients such as iron from great depths—often more than 1,000 meters (3,300 ft.)—where they hunt cephalopods, but they release them when they defecate near the surface. In many parts of the oceans, such as the Southern Ocean, iron is the limiting nutrient (meaning that its scarcity determines how many organisms can survive), which has led to the consideration of iron fertilization as a geo-engineering method to boost CO_2 fixation and sequestration.

Analyses reported in 2010 found that the only 12,000 sperm whales remaining in the Southern Ocean surprisingly make positive contribution to fighting climate change when they bring up nutrients from the deep and fertilize the upper layers of the ocean. They excrete around 50 metric tons (55 tons) of iron per year, 10 times more than researchers used in the LOHAFEX experiment conducted in 2009 to study iron fertilization. This iron causes the sequestration of 400,000 metric tons (441,000 tons) of carbon, or twice as much as the whales release through their respiration. Thus, working toward saving and expanding the remaining whale populations could be a significant contribution to fighting climate change and ocean acidification.

In a mathematical model investigating nutrient transport uphill from the deep sea to the mountain ranges, Christopher Doughty from Oxford University, United Kingdom, and colleagues quantified the capacity of whales to redistribute nutrients within the oceans, both horizontally along gradients and vertically against gravity.[1]

For lateral distribution, the authors estimated that whales are more efficient nutrient distributors than land mammals. However, due to large population reductions, their capacity has reduced by factors of 50 in the Southern Ocean and 7 in the North Atlantic.

For vertical fluxes, Doughty and colleagues used phosphorus as a proxy for nutrients that are often limiting in the oceans, including nitrogen and iron. Looking at nine species of marine mammals that are known to dive deeper than 100 meters (330 ft.), such as the blue whale (*Balaenoptera musculus*), sperm whale (*Physeter macrocephalus*), and humpback whale (*Megaptera novaeangliae*), they found that the whales' overall capacity of carrying phosphorus from the depths to the surface has fallen to less than a quarter of what it once was, from 340,000 metric tons (375,000 tons) per year to 75,000 metric tons (83,000 tons). Regional differences again suggest that the Southern Ocean suffered more severe losses than the North Atlantic.

These capacity losses are mainly due to large-scale whale hunting in the last three centuries. Although none of the great whale species has been hunted to extinction yet, many have suffered population losses of the order of 90%, and the marine megafauna is currently at risk of following its terrestrial counterpart into oblivion (see chapter 6).

If whales can revert the sinking of nutrients down the water column, seabirds can equally work against the flow by feeding at sea and leaving their leftovers on land, as the example of guano islands shows. The use of guano in agriculture has made this movement more rapid than it would normally be. As many species of seabirds tend to settle and defecate on bare rocks, away from the habitats of any predators that may be interested in their nests, their excrements may not be of immediate benefit to the terrestrial ecosystems. On longer timescales, however, they can get washed down to the vegetated areas. Even the guano islands could become reconnected with the land at low water levels, and the guano deposits could thus provide natural fertilizer for terrestrial ecosystems.

Doughty and colleagues estimate that 20% of the guano produced by seabirds drops on land. This corresponds to a total annual transfer of phosphorus from sea to land of 6,300 metric tons (6,900 tons) per year.

The contribution of seabirds is dwarfed by a less visible flux, provided by migrating fish species (anadromous fish) such as salmon, which migrate from oceans to rivers to breed and typically die there. Out of an estimated 110 extant species with this life cycle, the researchers studied the distribution of 42 species and calculated that they may have historically shifted 140,000 metric tons (154,000 tons) of phosphorus from sea to land. Severe population crashes have left only 4% of this carrying capacity intact. Thus, the authors estimate the current capacity of fish migrations to be 5,600 metric tons (6,200 tons) per year, which is comparable to that of seabirds.

Finally, fishing by humans also retrieves nutrients from sea to land. At a total of just over 100 million metric tons (110 million tons) of fish captured per year, the fisheries industry is now the largest importer of nutrients from sea to land—although it tends to be highly unsustainable in its target selection and thus may ultimately do the redistribution of nutrients more harm than good.

On land, nutrients are distributed unevenly and always remain at risk of being washed down to the sea. Distributing them to areas where they are less abundant and against the hydrological flow is thus an important ecosystem service. In their previous work, Doughty and colleagues had shown that megafauna makes a disproportionately large contribution to this nutrient distribution process, based on their wider-ranging movements.

Thus, the almost complete loss of the Pleistocene terrestrial megafauna on all continents except Africa (as discussed in chapter 5) has led to a collapse of the capacity for lateral nutrient transfer on land. In the Amazon basin, for instance, this leads to a drastically reduced transfer of phosphorus away from the fertile river plains and toward uphill areas where it is in short supply.

Even Siberia was a more fertile land than it is today when it still hosted woolly mammoths. Beyond their fertilization services, Pleistocene proboscideans, the taxonomic order from which only the elephants survive, were efficient ecosystem engineers that kept vegetation in check. Their extinction produced an ecological state shift that is discernible in the fossil record in areas like Beringia, the Pacific Northwest, and the Northeast of the United States, as Anthony Barnosky from the University of California, Berkeley, and colleagues reported.[2]

Doughty and colleagues calculated that the capacity for lateral nutrient transport on land has dropped to 8% of its Pleistocene level, due to the massive loss of megafauna.[3] Only exceptional areas with preserved megafauna, such as Kruger National Park in

South Africa, have stayed close to the original capacity. The largest loss of distribution capacity is observed in South America, where it dropped to 1%. This dramatic decrease is in part explained by the fact that, prior to the megafaunal extinction, South America was host to the largest populations of megaherbivores, all of which became extinct, and thus had the largest nutrient diffusion capacity to lose.

Connecting these links to form a chain, the researchers hypothesize that there once was a global animal-driven pump recovering nutrients from the deep sea and moving them upstream, even to barren mountain slopes, in partial compensation of the nutrient loss caused by hydrological flow and sedimentation. Overall, they calculate that the extinction of the terrestrial megafauna and the loss of marine mammal populations have reduced the carrying capacity of this natural nutrient pump to around 6% of what it was in the Pleistocene.

Moreover, the authors noted that "each step is potentially a non-linear positive feedback of increasing productivity." Thus the benefits to Pleistocene ecosystems may have been even larger than the pure numbers suggest. On the other hand, the losses brought on by the megafaunal extinction are equally likely to be larger than we can imagine.

Apart from the residual megafauna surviving mainly in Africa and South Asia, there is also a new kind of megafauna that has spread around the planet in the last 12,000 years, namely domesticated livestock. While the sheer biomass of our livestock has been estimated to exceed Pleistocene animals, our pigs and cattle are failing to fill in for the ecosystem services that mammoths and mastodons once provided. That is mostly because they are caged or fenced, so they are physically barred from transporting nutrients anywhere.

Doughty and colleagues argue that the current agricultural practice of having only one species in an enclosure and giving it very little space to roam ensures that the animals concentrate the nutrients rather than dispersing them. They suggest that mixed pastures or, better still, free-ranging herbivores could help to restore some of the nutrient transport capacity we lost in the megafaunal extinctions.

Instead, livestock are part of the highly artificial one-way nutrient flow of agriculture, whereby nutrients are taken from inorganic sources and end up as polluting waste. Nitrogen, for instance, is taken from the air, while phosphorus comes from rocks mined in Morocco, and both are used in large amounts as fertilizers. Nitrogen assimilation by the Haber-Bosch process has doubled the amount of nitrogen in circulation. Nitrogen compounds pose pollution problems in the aquatic system, and there are concerns about whether the natural nitrification processes that ultimately return nitrogen to the atmosphere can cope with the massively enhanced flow rates.

For phosphorus, it is not only a pollution problem but also a question of supply sustainability. Reserves of phosphorus minerals are found in very few places, and the only ones that will last for a few decades longer are in Morocco. Recovery and recycling of this nutrient must be developed with urgency to ensure supply security.

Considering these global cycles and flows, it appears that our civilization has been doing the exact opposite of the useful ecosystem service that megafauna used to supply to the global biosphere. While animals pumped nutrients uphill from the oceans to the mountains and helped to keep them in circulation, our current sewage systems are designed to dispose of them and let them go down the drain. Until a few decades ago, disposal just followed the "dilute and disperse" philosophy, whereby natural degradation processes in rivers and the sheer bulk of the water mass in the oceans were relied upon to take care of any pollutants and unpleasant smells.

Only after it became obvious that the sewage discharge had overwhelmed the natural capacity of rivers and killed off the life in many of them did the policy change to treating sewage before discharge into the water system, as is now common practice in the wealthier parts of the world. This has helped to improve the water quality in rivers, but it has done very little to restore nutrient cycling. Nitrogen may be returned to the air by denitrifying microbes, but other nutrients are still going down with the flow, unless they end up in the sewage sludge, which gets dried and either dumped or combusted. In all cases, most of the nutrients are lost rather than recycled.

Recovery of nutrients and other resources during wastewater treatment is only practiced in a few prototype plants. Using sewage directly to fertilize food crops has become unacceptable—not least because of the likelihood of finding drug residues and other organic pollutants, as well as toxic metals in the mixture. Using it to grow energy plants (e.g., sugar cane for bioethanol production), on the other hand, is one approach currently being tried. Producing biochar (a type of charcoal made from biomass) from combusted sludge or methane from anaerobic fermentation of sewage are other options investigated.

These trials have shown that well thought-out plans of resource recovery could turn treatment plants from a drain on the community finances into a source of income. More research and development are needed, along with the courage to try things differently. It's not always the best idea to go with the flow.

Now that we know that those big wild beasts were quite useful for the planet before they disappeared, what can we do to recover their ecological functions? Protecting whales and sharks while also saving elephants and rhinos from extinction would be a good start. Beyond that, re-introduction of free-roaming large herbivores like buffalo can be successful, as rewilding programs in Europe and North America have shown.

More adventurous spirits are already investigating the possibility of bringing back the mammoth to Siberia, where the last surviving specimens of this species died out only a few thousand years ago. The genome sequence of the extinct pachyderm was reported in 2008, opening speculation regarding the possible de-extinction of the species. This would naturally be a huge challenge and highly controversial on bioethics grounds, not to mention the religiously motivated objections to the creation of a species.

However, gene technology pioneer George Church from Harvard University, Massachusetts, has started working toward this goal, studying the genetic traits that made mammoths viable in the harsh conditions in Siberia. A full resurrection may still be off limits—not only because the mammoth DNA retrieved so far has been highly fragmented. Church also cites radiation damage accumulated over thousands of years as a key problem that would make cloning impossible even if a nucleus with seemingly well-preserved chromosomes were found. Church believes, however, that transfer of a couple of dozen mammoth genes into Asian elephants—which split from the mammoth line only six million years ago—could establish a high-latitude variant of mammoth-like elephants that could resettle Siberia.

Thanks to progress in gene editing using the CRISPR-Cas technology, inserting extra genes for cold adaptation into the elephant genome has become more straightforward. In early 2015, Church announced that the first insertion of mammoth genes into elephant cell cultures had been achieved.[4] In October 2015, Church told the website Strange Biology that his team had "made 15 changes in the elephant genome using CRISPR." He stated, "This would make a 'cold-resistant Asian Elephant.'"[5] Given the slow life cycle of the animals, it would take at least a decade before he could watch a fully grown Siberian elephant trample the tundra.

Meanwhile, in Siberia, Sergey Zimov, director of the Northeast Science Station in Cherskii in the Russian Republic of Sakha, has set up a Pleistocene Park to study the ecosystem services of megafauna and try to re-establish a Pleistocene-style population of large beasts. The project has already re-introduced or encouraged the expansion of several large animal species, including reindeer, moose, wild horses, and musk oxen, as well as predators, such as wolves, bears, lynxes, wolverines, and foxes, on a current area of around 160 square kilometers (62 mi.^{2}) (see https://pleistocenepark.ru/).

Zimov has already discussed the possible mammoth project with Church. His observations so far suggest that bringing back a mammoth-like proboscidean would not only enrich the landscape but also help to lock in the permafrost that stores large quantities of methane. This is because the large-footed heavyweights have a unique way of compacting the snow and the ground that helps to stabilize the tundra environment.

Thus, in Siberia, as in the oceans, large mammals can help to achieve what our own species blatantly fails to do—stabilize the climate.

Salmon face uphill struggle

Rivers flowing to the sea naturally carry everything movable, including biologically important nutrients, in the downhill direction, from the mountains to the plains, from the plains to the oceans, where valuable material might sink to the seafloor and get buried in sediment.

For life to thrive on land, it is vital that biological mechanisms oppose this trend and carry nutrients against the hydrological cycle. Whales play their part in this, as do seabirds, as discussed above, as well as bears. Generally megafauna makes a bigger contribution than smaller animals, which is one of the reasons human-made megafaunal extinctions are bad for the entire living planet.

A surprisingly large contribution, however, is that made by anadromous fish—those species, like salmon and sturgeon, that spawn in fresh water but mature in the ocean. As the fishes put on most of their body weight in the sea but tend to die near their spawning ground or on the way there, they essentially carry their own body's worth of biomass uphill, often over enormous distances and to impressive altitudes.

Atlantic salmon (*Salmo salar*) and several species of Pacific salmon (genus *Oncorhynchus*) share the anadromous lifestyle that has fascinated observers and researchers for centuries. After hatching in a gravel bed in a river, the early juvenile stages stay within their native freshwater environment for a few months, depending on the species. When they develop into so-called smolts, they undergo physiological changes enabling them to live in seawater. The smolts migrate downstream until they reach the ocean.

During their oceanic phase of at least one year and up to five years, salmon mature and gain body weight of up to 20 kilograms (44 lb.). In an annual mass migration, known as the salmon run, the mature fish migrate back upstream to their hatching grounds. Tracking and genetic studies have confirmed what folk wisdom has known for centuries—namely, that most salmon return to the same river where they were born and even to the same hatching ground.

Several studies have tested the advantages of returning home simply by transplanting mature salmon to an unfamiliar environment. Staying closer to the natural situation, Kenyon Mobley from the University of Helsinki, Finland, and colleagues used population genetics to determine the reproductive success of salmon that returned to their home ground versus those that spawned in unfamiliar territory.[6]

The researchers analyzed the DNA of thousands of individuals obtained at spawning grounds in the north of Finland over four years. The results showed that local males had three times more offspring than strays, while the home advantage for females was close to a factor of 10.

Much research effort has been dedicated to the navigation mechanisms that salmon use for their astonishing homing. Their sense of smell is important, and they are known to distinguish their home river by its smell. In addition, they possess a magnetic navigation system that enables them to use both the inclination angle and the total strength of the Earth's magnetic field. While both these parameters change with latitude, they change in subtly different ways and can thus provide independent parameters for a magnetic identification of geographic locations.

Michelle Scanlan from Oregon State University in Corvallis and colleagues demonstrated that the ability to respond to these magnetic parameters is maintained even in non-migrating populations of Atlantic salmon that have remained "landlocked" in freshwater lakes for many generations.[7] Specifically, the researchers studied a population that has been resident in a lake in Oregon for the last 60 years, and which before that had been living in Maine rivers genetically separated from migrating populations for around 9,000 years. Although it has undergone some genetic changes, this population has retained the ability to orient in response to magnetic map information. Importantly, the juvenile salmon in this study were navigationally naïve, so there was no process by which they could have learned to orient in magnetic fields, suggesting their response was innate.

The sheer scale of these upstream migrations can be astonishing. The Chinook salmon (*Oncorhynchus tshawytscha*), for instance, migrates up to 1,500 kilometers (930 mi.) up the Columbia River and into the Snake River in Idaho, ascending to more than 1,600 meters (5,200 ft.) altitude. Predators, such as North America's grizzly bears, are using the salmon run as a convenient food source. This is an important connection for the global nutrient cycles, as it helps to ensure that the nutrients carried against the flow stay up in the mountains for a while.

Salmon migrate hundreds of kilometers up their native rivers, facing threats from predators and human anglers. Beaver dams have been shown to be helpful to their migration, but the much larger dams that humans build to generate electricity are an insurmountable barrier. Even where fish ladders are installed to enable fish to bypass the dam, the populations of anadromous fish are often reduced.

A case that has been much debated is that of Snake River in Washington State, the biggest tributary of the Columbia River. Historically, the river was part of one of the world's longest and most productive salmon runs, seeing fish travel to spawning grounds as far inland as landlocked Idaho. Now this path is blocked by eight major dams, and interventions, such as bypass mechanisms for the smolts' downstream migration and fish ladders for the mature salmon traveling upstream, were needed to keep salmon stocks in the rivers.

Hydropower has been a major commercial success in the Columbia River basin, with more than 250 dams providing cheap electricity to the Pacific Northwest and even enabling the utilities to sell surplus energy to California. Now, however, as author Jacques Leslie argued in an opinion piece in the Yale University online magazine *Yale Environment 360*, the availability of cheaper solar energy means some of the dams could be made redundant.[8]

Specifically, Leslie made the case that the four dams on the lower Snake River are offering very little commercial interest, and the restoration of the natural flow in this river section could have major benefits to the salmon and the river's entire food web, from the bears inland to the orcas off the Pacific coast.

Lisa Crozier from the Northwest Fisheries Science Center in Seattle, Washington, argued, however, that the survival rates of Chinook salmon on this run have already much improved since the initial population crash after the dams were built. As of 2019, fishing along the lower Columbia River is the main cause of salmon mortality during the upstream migration. According to Crozier, "Mean survival through the Snake River hydrosystem is 95%, and as high as 97%–98% in environmentally favorable years."[9] Crozier's colleague James Faulkner and others published a detailed analysis of the survival rates on this salmon run.[10]

Even when they are not facing barriers to their migrations, salmon may be suffering from human activities in other ways. They tend to accumulate toxic metals like lead from their environment, which is also bad news for salmon eaters, from grizzlies to humans.

Degraded rivers tend to be poorer in dissolved oxygen, which the salmon eggs, in particular, need to survive. Jack Bloomer and colleagues at the University of Southampton, United Kingdom, showed that different populations of Atlantic salmon responded to different oxygen levels in their spawning grounds by adapting the permeability of the membranes surrounding their eggs.[11] Analyses using electron microscopy showed that the thickness and porosity of these proteinaceous fibrous layers vary between populations and determine the oxygen supply rate through the membrane. As the authors noted, this adaptation to the particular oxygen stress situation of their environment should be considered when salmon are bred or transplanted to boost stocks.

Another way in which salmon evolution responds to environmental change involves adjusting their age at maturity, which also affects their adult size. In a perfect habitat, salmon benefit from maturing over several years and growing to sizes of around 20 kilograms (44 lb.). These larger individuals generally have better reproductive success than younger, smaller fish. The evolutionary trade-off to be made here arises from the accumulating risk. If life is so dangerous that too few individuals survive for several years, the population survival benefits from an earlier maturity and reproduction.

Age at maturity in salmon is a parameter that is particularly well suited for studies of molecular evolution, as it is largely controlled by a single genetic locus, the area around the VGLL3 gene. Craig Primmer's group at the University of Helsinki, Finland, established the connection between the distribution of the relevant alleles of this gene and the trend toward earlier maturation in two populations of Atlantic salmon.[12] The researchers recorded a long-term decline of the late maturity allele in one population, where its prevalence decreased by 18% over 36 years, in line with the observed size change of adult fish arriving at the spawning grounds.

More generally, the "portfolio effect" of having different life-history traits within one species such that certain populations can better cope with certain changes has been discussed for various salmon species and is generally considered an advantage. By analyzing isotopes in salmon otoliths (ear stones), Sean Brennan at the University of Washington in Seattle and colleagues showed that a similar flexibility also applies to the use of the mosaic of habitats available across large watersheds in Alaska.[13]

Salmon populations will need their flexibility and resilience, as many of them are likely to suffer from both major effects of greenhouse gases emitted into the atmosphere—namely, climate warming and ocean acidification. In a comprehensive study of 33 population groups of Pacific salmon on the West Coast of the United States, Lisa Crozier and colleagues analyzed 20 different attributes to assess each group's vulnerability to environmental change. The populations ranked most vulnerable overall were Chinook salmon in the California Central Valley, coho (*Oncorhynchus kisutch*) in California and southern Oregon, sockeye (*O. nerka*) in the Snake River basin, and spring-run Chinook in the interior Columbia River and Willamette River basins.[14]

Ocean acidification, which is already known to endanger corals and other marine organisms, can also harm salmon populations. Chase Williams and Andrew Dittman, also at the Northwest Fisheries Science Center, together with Evan Gallagher from the University of Washington investigated the effect of increased CO_2 in the water on olfaction-dependent behaviors in coho salmon (*Oncorhynchus kisutch*). They discovered that CO_2 concentrations predicted to occur in seawater in the near future disrupted response pathways in the olfactory bulb and rendered the fish unable to react appropriately to danger signs, such as the presence of skin extracts in the water.[15]

Salmon has always been a popular fish to catch and eat, to the extent that there have already been numerous attempts to translocate them to non-native environments, and there is a large and growing industry of producing salmon either in aquaculture or by assisting the reproduction of wild populations.

Salmon aquaculture grew dramatically in the last decades of the twentieth century. In 1996 its production first surpassed the harvest of wild salmon. Typically, aquaculture reduces the life cycle to two stages. Salmon are hatched and grown in freshwater tanks at first and then transferred to large nets suspended in the ocean, where they are fed until they are harvested. The industry has grown particularly along the shores of Norway, Scotland, and Chile, where seawater with the appropriate temperature is available in fjords offering shelter from storms. However, aquaculture business has also sprung up in areas not usually known as salmon habitat, including New Zealand and the United Arab Emirates.

The high density of fish crowded in seawater pens has brought several serious problems, including the spread of diseases and parasites such as sea lice, which can also affect wild populations coming too close. Accumulation of heavy metals has also been reported for farmed fish, leading some environmentalists to label the products the most toxic food in the world. However, consumer associations continue to recommend farmed salmon products, and their growth appears to be unstoppable.

Alaska banned aquaculture of finfish including salmonids in 1990. Hatcheries are still being used there in an approach that is referred to as ranching, as opposed to farming. The key difference is that the fish swim freely during their oceanic phase.

After a long tradition of exploiting salmon in many of its habitats, the recent shift to hatcheries and aquaculture has dramatically changed the nature of the interaction between humans and the widely appreciated fish, turning them into another industrial commodity.

In the interest of biodiversity and nature conservation, there are still lessons that can and should be learned from traditional ways of managing and exploiting salmon. Based on their study of how well traditional knowledge of the Sámi people in Finland and Norway is integrated with science and modern fisheries management, Camilla Brattland and Tero Mustonen from the University of Eastern Finland concluded that many aspects of this knowledge transfer could still be improved.[16]

The authors highlighted as a role model to follow, the Näätämö co-management project, a successful salmon rewilding project launched in 2011 on the Näätämö river in northeastern Finland. The project is led by local representatives of the Sámi, Indigenous reindeer herders living in the Arctic parts of Norway, Sweden, Finland, and western Russia. Salmon fishing is an important part of their traditional lifestyle.

Key parts of the project include restoration of spawning areas and continuous monitoring of the ecosystem health by the local Sámi people. Several rivers and Lake Sevettijärvi are also being restored. The hope is that the project can set an example to show how humans can exploit a wild species without having to use industrialized methods, working with nature rather than against it.

Hard times for ecosystem cleaners

El condor pasa—the soaring notes of the most famous of all South American folk melodies evokes the effortless flight of one of the largest birds that take to the air. Across the continent, the Andean condor serves as a cultural icon in all sorts of contexts. It graces the coats of arms of several countries and the logos of several airlines and travel companies, and it appears in countless images, myths, and legends.

The condor was once spread along the entire length of the Andean mountain range, from Venezuela to Patagonia. It has now become rare in Venezuela and Colombia. Because of this decline and because its long life span and slow reproduction rate make it vulnerable, it is listed as near threatened, but overall it is doing much better than most of the group it belongs to. As the binomial name (*Vultur gryphus*) conferred to it by Linnaeus indicates, the iconic bird is a vulture.

Phylogenetic studies suggest vultures are not a monophyletic group (i.e., comprising all and only descendants of a common ancestor), although the exact relationship between Old World and New World vultures remains to be established. Represented as a foraging guild defined by their activity as scavengers, vultures share crucial traits adapted to their role in breaking up cadavers, including some presumed to arise from convergent evolution.

Animal carcasses rotting in the open can become a breeding ground for pests and pathogens, can poison water supplies for other animals and humans, and can encourage opportunistic but inefficient scavengers like rats and feral dogs. Vultures are specifically adapted to facilitate the removal of carcasses. By breaking them up with their beaks, vultures enable access for other scavengers. In the African savanna, a well-defined division of tasks exists between several species of vultures specializing on different body parts, up to and including bones being consumed by the bearded vulture (*Gypaetus barbatus*). With their extremely acidic stomachs, they are better protected from pathogens they may find in carcasses and are thus able to remove them permanently.

In a review of the literature on pathogens found in vultures, Pablo Plaza at the National University of Comahue in Bariloche, Argentina, and colleagues concluded that, while vultures are often found to carry pathogens, there is no evidence that they risk spreading them to humans.[17] On the contrary, the data suggests that their role in cleaning up carcasses contributes to a healthier environment for other species and thus reduces infection risks for humans as well.

These services are now under threat, as 9 of the 22 existing vulture species are critically endangered, and a further three species are endangered. Dramatic declines have been observed, for instance, in India. There, the use of the anti-inflammatory drug diclofenac in species that the vultures feed on has led to population declines of over 97% for native vulture species, including the Indian long-billed vulture (*Gyps bengalensis*) and the red-headed vulture (*Sarcogyps calvus*) in the 1990s.

India banned the veterinarian use of diclofenac in 2006, but the vulture populations remain severely limited. They are still at risk where the drug is used illegally, as even small exposure is lethal for some of the more vulnerable species. Recovery can only proceed very slowly, as vultures have a long life span and slow reproduction rate.

The consequences of the lack of vulture services have become all too obvious. Large populations of cattle left to die naturally in the Hindu traditions were normally disposed of by vultures. After the population decline, less efficient mammalian scavengers, including rats and feral dogs, have moved in. Unlike the vultures, these mammals are likely to spread diseases such as rabies, anthrax, and plague to other animals and to humans. In India, the spread of rabies by scavenging feral dogs has become a significant problem, with 30,000 people dying from rabies annually.

The decline has also affected the Zoroastrian community in India, the Parsi, who have had to change their funeral rites established for more than two millennia. Traditionally, they consigned their dead to "Towers of Silence," where their bones were cleaned up by vultures within a few days. After the decline in India's vultures, the Parsi had to install solar reflection panels to help decomposition; some Parsi have accepted cremation, which orthodox Zoroastrians consider a sacrilege.

The consequences of vulture declines are also felt elsewhere. In 2020, in the hot desert of Sonora, northern Mexico, an outbreak of hemorrhagic disease caused many deaths in wild rabbit and jackrabbit populations. As vultures were slow in consuming these cadavers, a noticeable smell of decay spread in the landscape.

Conservation experts fear that African vultures may follow the Indian ones to the brink of extinction and cause even larger ecological ripple effects across the continent. Four species were classified as critically endangered in 2015. Intentional and accidental poisoning is a leading cause of vulture declines in Africa. When farmers use poison to fight carnivores, vultures may become collateral victims. Worse still, there have been reports of poachers deliberately poisoning vultures to cover their crimes, for fear the swooping scavengers might alert law enforcement to their activities.

An international collaboration set up by the University of Maryland with conservation experts and local communities in South Africa aims to address the causes of unnatural vulture deaths, poisonings in particular. In 2020, the team reported a novel set of tactics and strategies to prevent poisonings based on established knowledge from criminology that hadn't previously been applied in conservation science, using the Great Limpopo Transfrontier Conservation Area in South Africa, where poisoning has become a problem, as a model site.[18]

Raising awareness in the local communities of the dangers that poisons pose both for the local ecosystems and for the people who use them is one of the strategies that co-organizer of the initiative André Botha from the Vultures for Africa program at the Endangered Wildlife Trust at Midrand, South Africa, and co-workers are already implementing on the ground. They found that many local farmers didn't know it was illegal to poison wildlife, nor that it could harm vultures and endanger their ecosystem services.[19]

Elsewhere in Africa, a wave of vulture deaths was observed in Guinea-Bissau in February 2020. Overall, more than 2,000 vultures died under suspicious circumstances. Most of these were critically endangered hooded vultures (*Necrosyrtes monachus*), of which Guinea-Bissau hosts around one-fifth of the surviving global population. As there was no big-game poaching problem in Guinea-Bissau, experts suspected that the motivation was to harvest vulture body parts for belief-based practices, which are still widespread in West Africa. This interpretation was confirmed by the finding that many of the dead vultures were beheaded.

Preliminary toxicological analyses reported by the Vulture Conservation Foundation found highly toxic pesticides of the carbamate type. These are agricultural pesticides frequently used illegally to poison wildlife in Africa.[20]

Illegal trade of vulture parts is also a recognized problem in Nigeria. In May 2019, the Nigerian Conservation Foundation in partnership with BirdLife Africa and with funding from the United States Fish and Wildlife Service launched a project aiming to address this complex issue at its source. The work includes raising awareness among traditional medicine practitioners of the herbal alternatives to vulture parts, as well as measures to improve enforcement of the relevant laws.[21]

Apart from poisoning and poaching, other anthropogenic factors contribute to the decline of African vultures. Due to their large size, vultures are prone to die from collisions with wind turbines not built to safe design standards and from electric power lines.

These issues were addressed in the 166-page strong Vulture Multi-species Action Plan (Vulture MsAP), which was adopted by member states of the Convention on the Conservation of Migratory Species of Wild Animals (CMS) in 2017, and which provided a comprehensive, strategic conservation plan covering the geographic ranges (128 states) of all 15 species of migratory African-Eurasian vultures and promoted concerted, collaborative, and co-ordinated international actions.[22]

In February 2020, the CMS released the first version of the Vulture MsAP Strategic Implementation Plan, which includes an inventory of vulture conservation activities to date and provides a roadmap of concrete priorities for delivering the agreed framework of the Vulture MsAP.[23]

An important aspect of vulture conservation efforts that still needs improvement is international co-operation across the regions covered by the threatened species. As vultures cover vast areas and some species migrate long distances, local action needs to be backed up by efficient co-operation.

Birds that range widely and migrate over large distances, as many vultures do, pose challenges for conservation efforts. They are likely to face different threats in different places, and conservation work may not be possible where it is needed. On the plus side, however, recent progress in tracking technology has made it relatively easy to record the migrations of such large birds.

Evan Buechley from the University of Utah in Salt Lake City and colleagues used GPS tracking devices to follow the movements of 45 Egyptian vultures (*Neophron percnopterus*) during a total of 75 migrations recorded between 2010 and 2017 at the crossroads between Europe, Asia, and Africa.[24] The area known in bird migration studies as the Red Sea Flyway is the second-busiest flyway globally, surpassed only by the Pan-American route.

Around half the global population of this endangered species passes through this flyway. Buechley and colleagues addressed the question of where the bottlenecks and the busiest areas along the route are, with a view to identify conservation priorities. Any dangers awaiting the birds at a bottleneck of their route will affect a larger number of individuals, so improvements made there could be more effective than in areas where the populations are more dispersed.

Vultures use thermals to gain height and then glide to the next one. Therefore, they tend to stay close to land, where they are more likely to find suitable thermals. Accordingly, the researchers found major bottlenecks of the vultures' migration paths at both ends of the Red Sea, where the crossing into Africa can be made without flying over water for too long. The routes, visiting a total of 38 countries, also bundled along both coasts of the Red Sea and through the Levant and Anatolia but otherwise dispersed toward a variety of destinations at both ends of the migrations.

Due to the political instability in the area, large parts of the Red Sea Flyway are less than ideal territory for conservationists to work in. However, the bottleneck at the southern exit of the Red Sea, the Bab-el-Mandeb Strait, touches the small state of Djibouti on the African side, which is considered a safe and stable place where conservationists could conceivably conduct in-depth studies on the birds passing through.

The longest migration recorded in the Red Sea project spanned 12,000 kilometers (7,500 mi.), and the vultures typically travel 360 kilometers (225 mi.) in a day.[25] A better understanding of how large birds can achieve this kind of long-haul travels comes from South America.

Flight powered by flapping wings becomes increasingly uneconomical with increasing body weight, which is why larger birds make use of thermals and even of the drag created by other birds when flying in formation. To explore the physical limits of soaring, Emily Shepard from Swansea University, United Kingdom, and Sergio Lambertucci at the National University of Comahue in Bariloche, Argentina, and colleagues attached custom-built flight loggers to eight juvenile Andean condors to observe their flight behavior.[26]

With devices capable of logging every single flapping of the animals' wings, the researchers established that even the young, inexperienced condors made optimal use of thermals, essentially limiting the use of muscle-powered flight to takeoffs. The data show that wing flapping occurred only 1.3% of the recorded flight time. One condor flew 170 kilometers (106 mi.) in five hours without flapping its wings once.

The authors note high mountains of the Andes generally produce more upwinds than other terrain, possibly explaining the eastern range limits of the condor. They also discuss the implications of their findings for the flight physics of even larger now-extinct bird species. Comparisons with marine species such as the equally large albatross may be more difficult, however, as these face different atmospheric conditions and may use different flight mechanics.

Studying the condor's soaring lifestyle also casts an interesting light on the energy balance of vultures more generally. Soaring and gliding is basically free, with energy expenditure only slightly higher than at rest. Takeoffs, by contrast, are expensive, and spending time on the flat ground (as opposed to the inaccessible rock perches where they tend to nest) exposes the birds to the risk of attacks by mammalian predators.

Therefore, a condor must choose wisely where it lands. Studies using experimental placements of carcasses have shown that the birds will disregard food opportunities in certain locations. Majestically staying aloft and on the lookout may be the trait that has earned the condor its great popularity. As we depend on their ecosystem services, it would be wise if we could extend our affection for the soaring condor to other vultures.

CHAPTER 10

Life in the Times of Climate Change

Among the many ways in which human activities damage the biosphere and the ecological balance of nature, climate change has received a large part of the attention, although unfortunately not in the way it deserves. Considering the science, it has been clear since the 1990s that humanity needs to stop burning fossil fuels on the scale we are used to. We had 30 years to change our ways, but we didn't.

What happened instead was that we rolled out our Western, fossil-fuel-intensive lifestyle to the rest of the world. The global population of private cars passed one billion in 2010 and is projected to reach two billion in 2030. According to the eponymous UK pop song, there were nine million bicycles in Beijing at the beginning of the twenty-first century—now you'll be more likely to find the equivalent number of cars there.

As the 26 UN Climate Change Conferences (Conferences of the Parties, or COPs) came and went, the famous hockey-stick graph of atmospheric carbon dioxide measured in Hawaii continued to rise relentlessly, without the slightest sign of a slowdown. In fact, annual emissions kept rising, with the COVID-19 pandemic only making a small dent.

There are many other books on climate change and what needs to be done, including Gaia Vince's 2022 book, *Nomad Century*, which I discuss below. Details of the ripple effects have also been discussed in other chapters in this book. So this chapter on climate is just a very short reminder of where we are in the wider context of the damage being done to all Earth systems and the ecological network of the global biosphere, and how some vulnerable species can already act as sensors warning us of the bigger catastrophe ahead.

As of early 2023, the damage is beginning to hit home. Reports of climate disasters are no longer focused on cyclones and coral reefs in the South Pacific. Impacts are also reaching the Global North, where most of the excess carbon dioxide came from. One might expect this to be the turning point where the big-emitter countries start to act, but don't hold your breath.

Arctic meltdown

In July 1893, explorer Fridtjof Nansen (1861–1930) and his crew left Norway to sail along the Siberian coast to the New Siberian Islands, then turn to the north. Nansen's plan was to let the purpose-built ship *Fram* be locked into the pack ice and move with the ice drift across the North Pole to emerge into the Atlantic Ocean on the other side. Although the plan was highly controversial at the time, it was backed by the evidence of debris from shipwrecks that had been carried from Siberia to the coast of Greenland.

The ship and crew successfully completed a three-year drift across the Arctic and emerged into the North Atlantic in August 1896. The ship missed the North Pole by just under five degrees (550 km, or 342 mi.), and Nansen's attempt to reach it on skis with just one crew member, Hjalmar Johansen, failed, though they set a new latitude record for explorers, reaching N 86°14′.

Despite not reaching the North Pole, the expedition counts as a major success and a milestone in polar exploration. It significantly advanced the knowledge of the Arctic by proving, among other things, that there was no major new landmass waiting to be discovered under the polar ice cap. If there had been land, it would have blocked the ice drift that carried the ship, and soundings taken along the journey consistently showed deep water across the Arctic.

Around 125 years after the *Fram*, another research vessel, the German research icebreaker *Polarstern*, which is operated by the Alfred Wegener Institute, was hitched to the polar ice to drift across the Arctic. Researchers on board found the Arctic environment dramatically changed since Nansen's days.

In September 2019, the *Polarstern* left Tromsø, Norway, to spend a year drifting through the Arctic Ocean, essentially following the ice drift much like Nansen's *Fram* did. Its voyage served an international project, named MOSAiC (Multidisciplinary drifting Observatory for the Study of Arctic Climate), jointly operated by 20 nations and funded to the tune of 150 million euros.

Whereas Nansen and his colleagues remained cut off from the world and any sort of comfort for three years, modern Arctic explorers can rely on hot showers, satellite communications, and helicopter transport. With research teams coming and going, more than 400 people took part in the MOSAiC expedition. Their data will keep them busy for years and will continue to inform climate science for decades.

What has been clear from the beginning, however, was that they traveled a dramatically warmer Arctic than Nansen did. Rather than being locked in and at risk of being crushed by the ice, the researchers simply tied their ship to a large block of ice, set up research stations on the ice, and followed its drift. In the summer, when COVID-19 restrictions forced the cancellations of flights that should have brought a new shift of researchers, the ship could just leave the ice floe to go pick up the researchers and come back a few weeks later. After *Polarstern* reached open water at the end of the drifting voyage, the expedition returned to the North Pole to examine a second ice floe at the onset of its second winter, before returning to its home port of Bremerhaven, Germany.

Climate science has long known that the Arctic is heating up much faster than the rest of the planet, but comparison of the new measurements with those that Nansen took 125 years earlier still came as a shock. "We've seen how the Arctic ice is dying. In the summer, even at the North Pole, it was characterised by extensive melting and erosion," said expedition leader Markus Rex from the Alfred Wegener Institute. "Though today the Central Arctic remains a fascinating, frozen landscape in winter, the ice is only half as thick as it was 40 years ago, and the winter temperatures we encountered were nearly always ten degrees warmer than what Fridtjof Nansen experienced on his ground-breaking Arctic expedition over 125 years ago."[1]

The detailed measurements, taken over the course of more than a year and covering an area around the North Pole that was inaccessible until very recently, will help to improve climate models and inform climate policy. It is to be hoped that the news of a dying Arctic will serve as a wake-up call to the world of politics to get policy efforts moving. "If we don't make immediate and sweeping efforts to combat climate warming, we'll soon see ice-free Arctic summers, which will have incalculable repercussions for our own weather and climate," Rex said.

These warnings came at the end of a year that had seen no shortage of climate records and warnings of dramatic consequences, especially in the Arctic. Although the Arctic Ocean shares currents with the Atlantic, the fact that it warms much more rapidly than the adjacent oceans appears to suggest that much of its climate problem originates on the continents that surround it.

Rivers, for example, not only deliver freshwater to the ocean—they also carry heat. In the case of the Arctic Ocean and the ice-melting dynamic, this heat transfer has long been neglected. Hotaek Park from the Japan Agency for Marine-Earth Science and Technology in Yokosuka, Japan, and colleagues quantified this contribution by combining existing modeling approaches for continental, sea, and ice temperatures and heat flows.[2]

The researchers found that increasing heat carried by rivers into the Arctic Ocean, especially along the Siberian coast, makes a significant contribution to ice loss. They estimate that it has caused 10% of the ice surface lost since the 1980s. Still, only a small part of the river heat goes into the melting. A larger part escapes to the atmosphere and continues to contribute to the general warming trends.

The warming conditions on the Siberian coast are particularly crucial because this is where much of the new ice is formed, which then travels on the transpolar drift toward Greenland. After a summer of record-breaking heat in Siberia and Scandinavia, the formation of new ice in the Laptev Sea got going at the end of October 2020, around 10 days later than the latest freezing periods observed previously.

On the other side of the Arctic Ocean, northern parts of the North American continent have also experienced exceptionally high temperatures. In October 2020, researchers from the University of Alaska Fairbanks, using the research vessel *Norseman II*, found that the ice formation in the Bering Sea was also delayed by weeks. And in July of that year, the Milne Ice Shelf on the coast of Ellesmere Island collapsed and started to disintegrate, losing 80 square kilometers (31 mi.[2]) of floating ice surface.

The loss of sea ice and floating ice shelves does not affect the sea level, as the ice displaces water according to its weight. The danger is, however, that the feedback loops caused by effects such as the loss of buttressing coastal shelves and the increased absorption of sunlight by open water replacing reflective ice will also destabilize the terrestrial ice caps like the one covering most of Greenland. Its loss would cause sea levels to rise by six meters (20 ft.).

The dramatic loss of solid surfaces not only on the Arctic Ocean but also on former permafrost soils around it is bound to change geography for many animals at high northern latitudes. It is therefore crucial to understand their movements to be able to assess how the loss of sea ice and permafrost ground will affect them.

Sarah Davidson at the Max Planck Institute of Animal Behavior in Radolfzell, Germany, and colleagues have therefore set up a new databank for movement data, called the Arctic Animal Movement Archive, which is hosted on the Max Planck platform, Movebank.[3] The datasets already present at the launch included millions of data points on 96 Arctic species. In their initial report, the researchers used the data to analyze changes in the migration behavior of the caribou (*Rangifer tarandus*; known as the reindeer in Europe) and the golden eagle (*Aquila chrysaetos*).

Canada's caribou populations are known for their extremely wide-ranging migrations coupled to their reproductive cycle. Some migrate hundreds of kilometers to reach specific calving grounds. As the snow and soil thaw earlier, the soft ground makes it harder for the animals to travel. The most northerly populations of caribou have responded by shifting their reproductive calendar—they now give birth significantly earlier in the year. For the golden eagle, the researchers reported that the migration toward breeding grounds in the north is affected only in immature birds, while the adult eagles stick to the established timing.

Marine animal species will also have to adapt to a warming Arctic. As the ice disappears, researchers are beginning to diagnose an "Atlantification" of the Arctic Ocean, meaning the marine environment will be more like the North Atlantic than the way the Arctic used to be.

There is also a greening effect, with primary producers being encouraged by the thaw both on land and in the water. Logan Berner from Northern Arizona University in Flagstaff and colleagues used satellite observations to analyze the vegetation changes in the tundra around the Arctic.[4] Between 1985 and 2016, data for North America and Europe showed an increase in the vegetation density (greening) in 38% of plots analyzed and a decrease (browning) in only 3%. Data for the entire Arctic including Siberia compared the years 2016 and 2000 and saw greening in 22% of locations versus browning in 4%.

Vegetation is also increasing in the Arctic waters. Based on satellite and ship-based measurements of chlorophyll density, analyzed with new algorithms developed specifically for the situation in Arctic waters, Kevin Arrigo at Stanford University, California, and colleagues calculated that the net primary production in the Arctic Ocean increased by 57% between 1998 and 2018.[5] The researchers estimated that, between 2008 and 2018, this increase was no longer driven by the decreasing sea-ice cover; rather, the change was mainly a result of increasing phytoplankton density in the waters, which had been free of ice since 2008. This may be related to the increased availability of nutrients coming in from the Atlantic. While greening may take up additional carbon dioxide in the short term, it is no remedy against climate change. This is because it also reduces the surface albedo (reflectivity) and thereby adds to the warming, threatening a feedback loop that might accelerate change.

Apart from the albedo effect, there is a potentially much bigger feedback loop to worry about. The warming Arctic is set to release large amounts of methane, both from the thawing permafrost soils and from methane hydrate deposits in the ocean. Preliminary observations from an ongoing research expedition led by Igor Semiletov of the Russian Academy of Sciences on board the research vessel *Akademik Keldysh* suggested that methane release from the continental shelf just north of the Laptev Sea (where ice formation was delayed by 10 days in 2020, as discussed above) was already underway. The researchers suggested that the Atlantification of the waters along the Siberian coast may have destabilized the methane hydrate deposits.[6]

Further feedback effects arise from human activities. As the ice disappears, new shipping routes become available and new opportunities for resource extraction emerge. The resulting pollution and disturbance are likely to further accelerate the decline of the Arctic as an ice-bound biome.

The Arctic is recognized as the fastest changing hotspot of the climate catastrophe. Just how fast things move up north was exemplified when researchers visiting the North Pole in 2018 left a time capsule. It contained objects documenting life in the early twenty-first century and letters intended for a future world in which the Arctic would be no longer frozen. Evidently, the researchers failed to consider the transpolar drift that carried both Nansen's *Fram* and the Alfred Wegener Institute's *Polarstern* across the Arctic. The time capsule was found on a beach in Ireland in October 2020, demonstrating once more that the ice-free future may be arriving much faster than anybody expected.

Global North feels the heat

On the prominent rock of Loreley on the Rhine River, a siren sitting on the edge of the cliff, brushing her golden hair and singing seductive songs allegedly distracted the shippers on the river to the extent that they often ran their ships aground. What the German poet Heinrich Heine (1797–1856) in his famous ballad from 1824 claimed to be a fairytale from ancient times was in fact a romantic invention that his colleague Clemens Brentano (1778–1842) made up just 23 years earlier. Both stories have shaped the view of the Rhine as having a romantic heritage at the heart of Europe, reinforced by the approximately 40 ruined or restored castles on the 65-kilometer (40 mi.) stretch around Loreley, the Upper Middle Rhine Valley, which was added to the UNESCO (United Nations Educational, Scientific and Cultural Organization) World Heritage List in 2002.

In the summer of 2022, the passage through the most romantic part of the Rhine became fraught with danger again, and singing sirens had nothing to do with it. At the beginning of August, low water levels forced shipping traffic to reduce loads and proceed with extreme caution. By mid-August, the water level at the town of Kaub had dropped to the critical low of 40 centimeters (16 in.), the threshold below which shipping ceases to be economically viable. It stayed below this level for a week before it started to recover. Historically, the lowest water levels tend to occur in October, so this low arrived exceptionally early in the year and just after a series of record-breaking heat waves across the European continent, which had led to droughts and wildfires on a scale described as exceptional but possibly a portent of things to come.

July 2022 brought heat waves and record temperatures. On July 19, the United Kingdom recorded an all-time high of 40.3°C (104.5°F) in Coningsby, Lincolnshire—1.6°C (2.9°F) above the previous record from 2019. New record temperature highs were also reached in many places across Europe and in China. On September 6, records spread across California, with Sacramento reaching 46.7°C (116°F). All-time highs were also surpassed in neighboring states of Oregon and Nevada. (Climatologist Maximiliano

Herrera was kept very busy logging new records on his Twitter feed @extremetemps.)

Heat waves can endanger lives, especially in areas where residents aren't used to hot summers. Climate change, urbanization, and aging populations all combine forces to make the direct health impact and mortality attributed to heat waves much worse. Cunrui Huang from Tsinghua University, China, and colleagues have demonstrated this connection for the changes China underwent between 1979 and 2020.[7] The number of deaths attributable to heat waves was recorded as an annual average of 3,679 in the 1980s, but it increased dramatically and reached a peak of 26,486 deaths in 2017. The authors conclude that "increased exposure to heat waves, population growth and aging are the main reasons for the temporal increase and spatial variation of attributable deaths in China."

Moreover, if people turn to the use of air conditioning to keep the heat out of their homes, the emissions of power generation still relying on fossil fuels may exacerbate the problem and create positive feedback loops. In parts of China, the increased demand for air conditioning has led to power shortfalls, especially in the Sichuan region, which depends on hydroelectric dams for 80% of its power.

After the heat in Europe in 2022 came the drought, with major rivers like the Loire and Thames running dry locally and aquifers becoming dangerously low. Ghosts from the past emerged as the waters receded, including dozens of ships from World War II in the lower Danube and ancient monuments in Spanish reservoirs.

The United Kingdom as well as nearly half the European Union countries and half of China suffered a severe drought, making official responses, such as water rationing, necessary. The European Drought Observatory, associated with the European Commission, estimated that the 2022 drought was the worst that Europe had faced within the last 500 years. Heat and drought will also impact the harvest in the regions affected. Summer harvests such as maize, soybean, and sunflower were reported to have suffered, whereas many wine growers expected a better-than-average year.

Beyond the economic impact, continued drought can also bring health problems, as has been shown with the spread of valley fever (coccidioidomycosis) in California and neighboring states. Named for its spread in California's Central Valley, the disease is caused by a fungus that resides in the soil and can spread with dust in drought conditions. While most people infected suffer only mild respiratory illness, around 1% of cases lead to serious illness that can be fatal.

Prolonged drought also increases the risk of landscape fires, which, although a natural phenomenon in many ecosystems, can become devastating if fuel is allowed to build up. Whereas California had experienced catastrophic fire seasons for several years up to 2022, this exceptionally dry summer brought fires to areas of Europe previously spared. Fires were worse than before in the South of France and parts of Spain. Areas of Saxony, Germany, also experienced large fires, which destroyed record-breaking amounts of forest. On the outskirts of London, a wildfire managed to destroy a few homes.

The science of attributing specific extreme weather events to causes related to anthropogenic climate change has seen rapid growth recently, with better data and improved modeling making it easier to make quantitative connections. While events such as tropical cyclones are more difficult to pin down, the recent heat waves are relatively straightforward. Temperatures in the United Kingdom exceeding 40°C (104°F) would have been "extremely unlikely or virtually impossible" in the pre-industrial Holocene climate, as attribution researcher Friederike Otto from Imperial

College London, United Kingdom, confirmed in media interviews after the event.[8]

Ben Clarke from Oxford University, United Kingdom, with Friederike Otto and others has reviewed the status of climate attribution efforts.[9] So far, more than 350 studies have quantitatively analyzed the role of climate change in more than 400 extreme weather events.

Reviewing this growing body of evidence along with 2021 *Sixth Assessment Report* of the Intergovernmental Panel on Climate Change (IPCC), the researchers found that "heat extremes have increased in likelihood and intensity worldwide due to climate change, with tens of thousands of deaths directly attributable." They caution that any such figures almost certainly underestimate the true impact of climate change, as data from many places most severely affected, especially in the Global South, are often lacking or difficult to access.

The authors indicate that further improving our understanding of causality links between CO_2 emissions and extreme weather impacts is crucial for predicting and mitigating the damage of further warming. To improve the chances of mitigating future impacts of climate change, they called for more systematic recording of extreme weather (especially in the Global South), better coverage of attribution in terms of a wider geographic range and a broader range of hazards studied, and a broader analysis of risk that covers other factors that may currently be more significant than the climate link. On the last issue, the Clarke and colleagues cite a recent study of famine in Madagascar, where the acute problem had not shown a clear climate signature yet but the identification of present vulnerabilities enabled the scientists to anticipate damages likely to be caused by future climate warming.[10]

A key problem in the global failure to tackle climate change is the geographic separation between cause and effect. Most of the excess carbon emissions produced in the last two centuries came from Europe, North America, and, more recently, China. The first noticeable climate impacts, such as sea-level rise and more devastating tropical storms, on the other hand, disproportionately affected the Global South, including Pacific islands and low-income countries unable to physically shield their residents from extreme weather events. This has been discussed in terms of climate justice and an obligation of industrialized countries to help those most affected.

Extreme weather events like the floods in Pakistan in September 2022 are continuing to cause numerous casualties and vast damage, but they are failing to dial up the speed of climate action. When climate impacts come home to roost and hit Europe or North America, however, there is the fleeting hope that things might change at the root of the problem.

Time is already running out to limit warming to 1.5°C (2.7°F), as agreed upon at the COP21 climate summit in Paris in 2015, considering the 2021 IPCC report already estimates the current warming to be 1.2°C (2.2°F).[11] Commitments to reach net-zero carbon emissions by 2050, as made at the COP26 climate summit in Glasgow in November 2021, are simultaneously too little to keep warming below 2°C (3.6°F) (never mind the more ambitious Paris target of 1.5°C) and at risk of not being kept in any meaningful way, except perhaps by inventive arithmetic with some extra dose of greenwashing.

Soon after the COP26 climate summit, the war in Ukraine and the natural-gas supply crisis pushed climate commitments off the agenda. The very same governments that made climate pledges in Glasgow were back at their old game of investing billions of dollars in new infrastructure to secure a future with fossil fuels.

Climate-linked disasters around the world and data from the IPCC assessments show that we have already passed the point where emission management alone could have protected us. It is now clear that the world must both reverse the trend on CO_2 emissions and simultaneously adapt to the impacts of climate change that are already inescapable, even under a best-case scenario.

Cities are going to feel the heat first, and some are already busy adapting to a much warmer world. The city of Paris, France, already operates a communal cooling system using the water of the river Seine as a heat exchange, which serves the Louvre Museum as well as the National Assembly. In July 2022, the administration signed contracts to also cover schools, hospitals, and metro (subway) stations by tripling the underground pipe network to a total length of 252 kilometers (157 mi.) by 2042. Many other cities are rethinking their urban designs, adding trees, or developing zero-carbon transport strategies.

Still, with the best efforts in mitigation and adaptation, many cities and regions may reach a point where human habitation and activity is no longer viable. In her 2022 book *Nomad Century*, science writer Gaia Vince argued that managed migration of billions of people will become necessary to avoid the worst climate impacts.

As climate problems can trigger or exacerbate social and political unrest, it is often hard to draw the line between political and climate refugees. Vince argues, however, that climate refugees, at a cumulative total of 50 million, already outnumber those fleeing political pressure. Rural areas in the south of Iraq, for instance, are currently becoming uninhabitable and forcing farmers to flee, overwhelming cities such as Basra. Mounting climate impacts will continue to drive up the number of displaced persons, which reached 100 million in 2020 (including those displaced due to political or religious persecution or war), after tripling in just 10 years.

As climate change continues to get worse toward the second half of the century, billions of people will be forced to desert their homelands in tropical regions that will no longer be habitable. With touching, if possibly unjustified, optimism, Vince suggests setting up an international, possibly UN-sponsored, mechanism that would recognize the right of people to move to safer grounds and offer them opportunities to settle there.

So far, however, political responses to migration, from Trump's wall on the US-Mexican border to the United Kingdom's attempt to relocate asylum seekers to Rwanda, hasn't given the impression that the migration challenges ahead will be managed well. Vince writes, "Climate change is in most cases survivable; it is our border policies that will kill people."[12]

The northern countries that drove industrialization and hence climate change are now beginning to feel both the weather fallout and the human cost. Adapting in a way that is also globally just is the biggest challenge of the twenty-first century, and addressing this challenge will require societies and political leadership to change course quickly.

Volatile climate stirs bird life cycle

The Paris Agreement, like many other documents, frames climate change in terms of the increase of global average temperatures, which the world community was trying to restrain to 2°C (3.6°F). The rise in the average temperature of atmosphere and surface waters has, by itself, some very important impacts on the biosphere. While some species stand to benefit, many are certain to lose. Many species, including corals, operate within a narrow temperature window of viability and are seriously threatened even by temperature changes within the Paris limits. In addition, shrinking ice caps and warming ocean waters both make sea levels rise, inevitably causing damage in coastal areas as well as habitat loss for species depending on sea ice.

And yet, the greatest danger may reside not in the 2°C (3.6°F) shift itself but in the increased variation that a warmer world is exposed to. As events in the first two decades of the twenty-first century have demonstrated, warmer ocean surface waters make extreme weather events more likely. The underlying physics is quite straightforward. Warmer water evaporates more readily, and more energy is available to feed tropical cyclones. Flooding and storm damage are already more common, but, as the climate systems are disrupted, ripple effects can also include droughts and heat waves.

And yet, science and public perception often tend to treat each extreme weather event as a one-off (e.g., the record heat wave or the most devastating storm). Due to their sporadic appearance and individual nature, these events are not as easily modeled as a uniform shift in average temperature. Therefore, science has been slow to work out what damage the erratic weather in a warmer world might do to the biosphere. Systematic analyses are beginning to emerge, however. After several studies of plants, and especially of crops, more recent work revealed the sensitivity of the life history of birds to extreme weather events.

The black-browed albatross (*Thalassarche melanophris*) breeds on islands off the Antarctic coast, and its life cycle has been closely studied for decades. The female lays a single egg in October, which hatches in December. Both parents care for the egg and nestling. Once the chick has fledged, all albatrosses will migrate to spend the winter in warmer climes, in southern Australia or Tasmania.

A colony of albatrosses at the Canyon des Sourcils Noirs on the Kerguelen Islands, part of the French Southern and Antarctic Lands, has been observed for 50 years. Data of over 4,000 individuals show that the survival and reproductive success of the birds varies with changes of sea surface temperatures, which appears to influence young, middle-aged, and older birds in different ways.

To tease apart these complex correlations, Deborah Pardo and colleagues from the Centre d'Etudes Biologiques de Chizé in Villiers-en-Bois, France, developed a computer model that builds on the comprehensive empirical datasets reaching back to 1978 and explicitly includes age information for the birds, as well as distinguishing between changes to the long-term average temperatures and the temperature variability (i.e., the likelihood of extreme events).[13]

The analysis showed that a decrease in population growth correlates well with both the frequency and amplitude of extreme changes in sea surface temperature. By contrast, the long-term average temperature of the surface water, which gradually increases due to climate change, shows a positive correlation with population growth, because the historical average of sea surface temperatures was lower than the optimal temperature for this species.

Within limits, this positive reaction to warming can compensate for the negative impact of the extreme temperature variations becoming more frequent under climate change. However, this benefit will cease as soon as the actual average temperature passes the optimum to which the albatross is adapted, as beyond that temperature, negative results of mean temperatures and variability will combine their deleterious effects. The remaining margin for the albatross is only 0.1°C (0.2°F), while the IPCC projects mean temperature increases of 1°C (1.8°F), so this reversal of fortunes appears inevitable.

As for the age distribution of survival chances, the research found that older birds are vulnerable to both the rising average and the extreme spikes of temperature, an effect that the authors compared to the increased mortality among older humans observed during heat waves. While younger albatrosses are more resilient to these threats, they are exposed to dangers from fishing gear when they follow long-line fishing boats.

Overall, the authors expect the population in the Kerguelen to decline further, a problem which may become more acute as temperatures rise past the optimum and variability continues to increase.

Meanwhile, John Wingfield and colleagues at the University of California, Davis, reviewed several case studies of other bird species caught out by weather extremes in the polar regions.[14] Snow petrels, for instance, spend their entire life in Antarctica and are not easily shocked by snow or cold winds. In one episode of unusual weather, however, snowfall in their breeding season followed by unusual warming to over 7°C (45°F) and abundant meltwater led to dramatic breeding failure. What counts as extreme varies from species to species—for this one, it was a mild breeze that pushed it over the edge.

Similarly, the authors described how a colony of Adélie penguins (*Pygoscelis adeliae*) on Petrel Island, Antarctica, suffered total reproductive failure in the summer of 2013–14. Researchers have related this unprecedented event to an unusually warm summer with rain and melting slow, which became deadly for penguin chicks adapted to cold, dry conditions.

From the Arctic, Wingfield and co-authors related an example of low temperatures and snow coverage in the spring of 2013, which delayed the migration and breeding of the Lapland longspur (*Calcarius lapponicus*) and Gambel's white-crowned sparrow (*Zonotrichia leucophrys gambelii*). These events, though not extreme by the standards of Arctic weather conditions, arrived at the wrong time for both species, causing them to abandon their territories, which ultimately resulted in complete reductive failure.

Wingfield and colleagues concluded that an extreme event that requires behavioral or physiological responses cannot be defined solely based on climate parameters. Instead, each species has its own set of sensitivities toward deviations from the long-term average conditions to which it has adapted in the course of its evolution.

The researchers suggested using the so-called allostatic load approach instead, which quantifies the potential of a population to resist climatic perturbations. When this flexibility reaches its limit, the life history switches to an emergency state, which safeguards survival but may sacrifice brood success, as seen in the Arctic birds that gave up their territory.

Moreover, the authors suggested distinguishing between reversible (e.g., behavioral) and irreversible changes that reroute development to a different phenotype that may cope better with the extreme condition but will suffer a cost upon return to normality. The behavioral flexibility is often mediated by hormones, which are increasingly being studied.

Sensitivity to climate variations can even differ for closely related species that share similar lifestyles in the same geographic area. Janet Gardner and colleagues from the Australian National University in Canberra have demonstrated this with a pair of coexisting insectivorous passerine species, the red-winged fairy wren (*Malurus elegans*) and the white-browed scrubwren (*Sericornis frontalis*). Based on observation data covering 37 years, the researchers found correlations with climate events that were both complex and different between the species.[15]

The analysis showed that survival rates in both species correlated more closely with temperature extremes than with average temperatures. Specifically, the number of cold, wet days in a given winter correlated with decreasing survival rates in all years but one. These effects were observed both within the season of the extreme events and in a longer-term repercussion in the following season.

Body size is an important parameter in the vulnerability of birds to temperature variation, as the different surface-to-mass ratio of smaller birds makes it more challenging to maintain their body temperature, while larger body sizes require more energy to maintain. In the red-winged fairy wren, but not in the scrubwren, the data showed a reduction of body size over the 37-year observation period, in line with Bergmann's rule. Named after the German scientist Carl Bergmann (1814–1865), the rule originally described size variations between species of the same genus, such as penguins, but it has also been used to predict reduction in body size as climate warms. The observed survival rates would have predicted an opposite selection pressure, however, as larger individuals have better chances. Thus some of the complexities of these observations remain unexplained for now.

In another report on a well-studied small passerine species, Pascal Marrot and colleagues from the University of Sherbrooke, Canada, analyzed the impact of weather extremes on the breeding behavior of blue tits (*Cyanistes caeruleus*) in the South of France. Specifically, the researchers looked at extreme rainfalls and hot and cold days, and they tried to correlate these weather extremes with the birds' laying date, clutch size, and survival, based on observation data from more than 25 years and on population modeling.[16]

The laying date is a crucial parameter for the breeding success of tits, as they must find large numbers of caterpillars to feed their brood. Thus, the three-week period of parental feeding should be aligned as closely as possible with the availability of suitable caterpillars.

The research found that changes in average temperature had little effect. However, the occurrence of hot days during the early feeding period could impact survival to an extent that it drives selection to early laying. This effect is readily explained via the reduced availability of caterpillars.

Extremely hot days arriving in the late stage of the feeding period, by contrast, appear to help survival. While a mechanism for this correlation has not yet been established, the authors speculated that a reduced activity of predators targeting the fledglings may be the cause.

Overall, the authors concluded, the increasing frequency of extreme climatic events could soon become a substantial threat to populations of common birds such as the blue tit. Previous studies have shown that the intensification of agriculture and loss of insect prey have already decimated the populations of many bird species that were common only a few decades ago.

While researchers generally agree that climate change and, in particular, ocean warming are already increasing the likelihood of extreme climatic events and will continue to do so, there is less agreement on how to define and characterize these events.

Part of the problem is the diversity of climatic conditions that could become extreme for one species or the other. While the increase of average temperature is easily recorded as a single number changing over time, the recording of extreme events needs careful consideration of definitions not only based on meteorology but also on biological consequences.

Many bird species, with their complex life cycles often involving seasonal migration and elaborate breeding protocols, are particularly vulnerable to climate variations and can serve as indicator species, like literal canaries in coal mines. We just have to look out for them to ensure that their vulnerabilities don't set them on the path to extinction.

Climate canaries

Like the historic and now proverbial canaries in the coal mines of old, birds in general are a uniquely useful bioindicator for environmental problems affecting wildlife. Many of them are highly visible and often observed by numerous amateur bird-watchers, who can now easily contribute to large-scale distribution studies. Many species are sensitive to change for a wide variety of reasons.

Many island species have already become extinct after the arrival of humans and other invasive species. Migratory birds can be particularly vulnerable to problems occurring anywhere along their path, while on the other hand, the stationary nature of their extensive nesting and brood-care procedures often means that it is hard for birds to escape arising dangers. As the collective human consciousness is still too slow in grasping the severity of the climate catastrophe, many bird species are already fighting for survival.

In September 2022, BirdLife International, which had just celebrated its centenary in June, released the fifth edition of its report, *State of the World's Birds*. It warned that the situation for birds continues to deteriorate globally, as "species are moving ever faster towards extinction."[17] The report found half of all known bird species in decline. In the previous edition, in 2018, the figure was 40%. The report noted that 180 bird species had already become extinct in the last few centuries, and one in eight of the surviving species is threatened with extinction.

The report highlighted climate change as one of the causative factors, as well as agricultural expansion and intensification, unsustainable logging, invasive alien species, and overexploitation. On climate, the report warned that the greatest declines have been observed in areas that also saw rapid warming. Moreover, one in four threatened species is believed to have been negatively impacted by climate change.

The National Audubon Society says it engages with 1.8 million members and the 45 million Americans who consider themselves bird lovers. It released a specific analysis of the dangers of climate change for birds in 2019, dramatically titled "Survival by Degrees: 389 Bird Species on the Brink."[18] The title refers to the finding that 64% (389 of 604) of bird species in North America are moderately or highly vulnerable to climate change. It also highlights the fact that climate-change mitigation could still avert a large part of those threats.

The Audubon report and the society's website also provide detailed model predictions of how the ranges of specific bird species are likely to change. Their assessments are based on 140 million bird observations and state-of-the-art climate models. For the lucky birds, the northern limit of their range may move outward as much or more than the southern edge retreats. In less fortunate cases, a species' range may shrink, or the altered conditions may reduce its survival chances and reproductive fitness.

For instance, the northern pintail (*Anas acuta*), from the family Anatidae (ducks, geese, and swans), is currently widespread around the Northern Hemisphere, but it could suffer range shifts under climate warming of up to 3°C (5.4°F), which is the global change forecast for the end of this century. While the winter range of this species stands to gain surface area, its summer range is predicted to shrink somewhat, although the overall assessment of the species' situation is described as stable.

By contrast, the piping plover (*Charadrius melodus*) is assessed as highly vulnerable, as it is projected to maintain only 13% of its summer range. The losses of 87% will not be compensated by an anticipated 59% range gain, as the species moves north. It is already considered endangered in all parts of its range.

Clark Rushing from Utah State University in Logan and colleagues studied the distribution of 32 bird species of eastern North America across 43 years and found a remarkable split between those species that remain in North America all year and those that migrate to the tropics in winter.[19] The resident birds and also those that migrated within the temperate climate zones were generally able to expand their range northward as the climate warmed up, without losing the equivalent area at the southern edge. By contrast, the tropical migrants faced shrinking habitat as they were unable to expand further north while losing range in the south. This trend matches the observation that tropical migrants in

North America have been declining while some residential species were doing well, and it suggests that tropical migrants need conservation attention.

Mountain birds may also have the option to move to higher altitudes to stay with the climate they are adapted to. As BirdLife International noted in its report, species in the Peruvian Andes and in tropical Australia have already moved to higher altitudes.[20] This may cause problems, as the space available will naturally shrink and run out as they move up.

Benjamin Freeman from the University of California, Davis, and colleagues described the recent warming causing habitat shifts on the slopes of the Cerro de Pantiacolla, which rises out of the Amazonian lowlands in southeastern Peru, as an "escalator to extinction" that will lead to "mountaintop extinctions of species that only live near mountain summits."[21] The authors concluded that "high-elevation species in the tropics are particularly vulnerable to climate change."

Climate change is also likely to upset the calendar of birds' lives. Those species that time their migrations by weather events, such as the arrival of spring weather, may get into difficulties when these times move. Thomas Lameris from the Netherlands Institute of Ecology in Wageningen and colleagues studied the 3,000-kilometer (1,860 mi.) migration of barnacle geese (*Branta leucopsis*) from the coasts of the North Sea to the Russian Arctic.[22] As the weather conditions in their wintering grounds don't correlate well with those in the Arctic, the geese can't adjust their departure date based on predicting what conditions they will find. Instead, the researchers established that the geese speed up their migration in response to earlier Arctic springs, by cutting short their usual stopovers when the conditions suggest mild weather ahead.

Although this behavioral response enables them to co-ordinate their arrival time with early spring dates, they don't quite manage the same for egg laying. In the earliest springs studied, the geese missed the optimal window of food availability and as a result had less healthy goslings.

As the Audubon Society's director of climate science, Brooke Bateman, noted in a story on the organization's website, the black-throated blue warbler (*Setophaga caerulescens*) now migrates a whole week earlier than it did 50 years ago.[23] Bateman also warned that, along with earlier springs, false springs followed by unseasonably cold weather are becoming more common in North America under climate change. Migratory birds responding to the first signs of spring may find themselves exposed to wintry conditions they are not equipped to handle and may die as a result.

While changes to distribution and migration are readily obvious thanks to the many bird-watchers keeping an eye on our feathered friends, climate change may also damage the fitness of those populations that stay in place and are exposed to higher temperatures or extreme weather events. These effects may be more subtle and are only beginning to be explored in specific case studies.

Justin Eastwood from Monash University, Australia, and colleagues studied nestlings of a small Australian songbird, the purple-crested fairy wren (*Malurus coronatus*), exposed to different ambient temperatures.[24] Nestlings in general are vulnerable to changing conditions, as they are immobile and dependent on parental care, thus unable to escape any environmental dangers.

The researchers found that higher air temperatures in dry conditions during the nestling period were correlated with shorter telomere lengths in early life. In birds as in humans, telomeres protecting the ends of the chromosomes shorten with each cell division, and their eventual loss leads to cell death, making them a molecular clock that predicts residual life expectancy. Wrens with shorter-than-usual telomeres are thus likely to die earlier and to have less reproductive success over their lifetime.

Although the birds may find evolutionary paths out of this problem, such as evolving longer telomeres to start with or changing their breeding calendar, the authors pointed out that these options may not be viable, and that the sublethal effects observed may in the long term lead to population declines. They concluded that "current extinction estimates based on direct extreme temperature mortality events, range shifts, or life-history changes may underestimate the negative impact of climate change on biodiversity." They suggest that time-delayed consequences of sublethal impacts such as telomere shortening should also be considered in estimates of the likely effects of climate change on population survival and biodiversity.

David López-Idiáquez from the University of Montpellier, France, and colleagues spotted another subtle change that may be heralding problems. They studied two Mediterranean subspecies of blue tits (*Cyanistes caeruleus*), one near the city of Montpellier, the other on the island of Corsica.[25] Based on measurements taken from 5,800 blue tits captured between 2005 and 2019, the researchers found that the coloring of the blue crest and the yellow chest had faded in both populations. In Corsica, the loss of coloration correlated with the change of ambient temperature at molt time, leading the researchers to hypothesize a connection to climate change. Because the colors serve as mating signals, there is the danger that the changes may affect reproductive fitness.

Genetic analyses did not find any changes to the color traits over the course of the study, hence López-Idiáquez concluded, "It is important to stress that this change is not genetic but plastic, one of the ways of adapting to new environmental conditions."[26]

Desert birds arguably live on the edge of survivability already, and they often start breeding in response to rainfall. In many cases, this means that they can't shift their breeding time away from the hottest time of the year. Therefore, Nicholas Pattinson from the University of Cape Town, South Africa, chose one such desert-dwelling population, the southern yellow-billed hornbill (*Tockus leucomelas*) of the Kalahari Desert in southern Africa, to analyze how its breeding success related to local temperature increases.[27]

The species has a remarkable breeding strategy that involves the female sealing the entrance to the nesting cavity from the inside, leaving only a small hole for air and food supplies delivered by the male. Because of this lockdown situation, predators don't play a significant role, and the brood survival mainly depends on food availability and environmental conditions. The breeding of this population at the Kuruman River Reserve was monitored from 2008 to 2019. During this time, success rates collapsed from an average of 1.1 live fledglings per attempt to 0.4.

Matching the breeding observations to weather data, Pattinson and colleagues found that the success rate was negatively correlated with the number of days on which the maximal air temperature exceeded a threshold of 34.5°C (94.1°F), which is defined by the observation that at this temperature 50% of male hornbills start panting to dissipate the heat. Although drought within the breeding period also showed a correlation, the high-temperature effect was upheld even in years without drought. The researchers conclude that "this suggests that global warming was likely the primary driver of the recent, rapid breeding success collapse."

The authors also determined a temperature limit of 35.7°C (96.3°F) above which all breeding attempts failed. Based on current climate trends, they calculated that by 2027 this maximum will be exceeded at their study site throughout the hornbill breeding season, meaning that by then the heat will make it impossible for the birds to reproduce there. Although the hornbill is currently still listed as least concern on the IUCN Red List, the collapse of breeding success in response to temperature rise could endanger its survival in parts of its current range.

As BirdLife International highlighted in its 2022 report, birds can act as highly visible bioindicators of wider problems affecting ecosystems and biodiversity more generally. The widespread declines and range shifts observed over the last years and decades are thus warning signs indicating the general biodiversity crisis that we are facing, which is partially caused by climate change and will separately have similarly apocalyptic consequences.

CHAPTER 11

Our Shared Burden of Disease

Zoonoses—diseases transmitted from animals to humans—have arguably shaped human history like nothing else. When hunters and gatherers first settled down and domesticated animals, they created a fundamental problem that we are still fighting today. Living closely with animals, they heightened the risk of transmission of infectious agents, and by living in ever larger communities, they provided host populations large enough for diseases to establish themselves indefinitely.

Today's childhood diseases, such as measles and chickenpox, originated this way. Typically, these zoonoses were much deadlier in the beginning, but over the course of millennia, they adapted to an ecological equilibrium that spares lives but makes transmission more likely. Lacking the relative immunity that the farmers developed, hunter-gatherer societies succumbed to the germs of the invading armies of farming nations as often as to their weapons, as Jared Diamond explained in his widely known book *Guns, Germs, and Steel*.[1]

Considering that our civilization created this problem and carried it around the globe, we are remarkably ill prepared to deal with new instances of disease transmission from animals to people. For the past seven decades, we have enjoyed the protection of antibiotics against bacterial disease, but we are now at risk of losing it—again, largely a problem our society has brought upon itself by using antibiotics unnecessarily and recklessly, which fosters antibiotic-resistant pathogen strains.

The situation is much worse with viruses. Established viral diseases can often be prevented by immunizations, but if an animal virus newly acquires the ability to spread within the human population, as HIV and Ebola did in the twentieth century and SARS-CoV-2 (the causative agent of COVID-19) did in 2019—and as some of the more than 100 variants of avian influenza might do at any time—we are unprepared.

To prevent outbreaks, we need a better understanding of the natural animal reservoirs of such diseases and of the interfaces through which they cross over into humans. As with most other problems we are facing, preventing the next pandemic boils down to better understanding ecology.

Pandemics past and present

The COVID-19 pandemic was the first to be monitored by molecular genetic methods from the beginning. Throughout the pandemic, we had science on our side, lending support that could almost appear miraculous and that helped to prevent many deaths. We were able to test ourselves for the presence of SARS-CoV-2, we could get vaccinated against it, and the authorities could monitor the spread of the disease based on wastewater analytics and spot outbreaks even before patients report to hospitals with serious symptoms.

The tragic outcome that more than seven million people died from this new disease within its first four years, and many more were left with long-term consequences, is largely due to a failure of translating this groundbreaking science into public health policy and into behavior change at the population level.[2] Advance warnings were available, but the "let's see how bad it gets" school of thinking prevailed, especially among populist leaders and populations cherishing individualism, and measures were only taken at the last moment, when they became inescapable in order to prevent the collapse of health care provision.

In the twenty-first century, we have all the science we need to fight or even prevent global pandemics like COVID-19, but the forces of anti-science are severely hindering the efforts. In the history of human disease, this is a novelty. Previous pandemics struck people who had no way of knowing what was going on or how they might protect themselves. Only now, with the molecular toolkits for ancient DNA studies and the extra motivation provided by the global health crisis caused by COVID-19, are scientists beginning to understand the major pandemics of the past.

The flu pandemic at the end of World War I ravaged countries already weakened by war and food shortages. Even though it killed an estimated 50 million people and proved surprisingly deadly to those aged 20 to 40, it was soon swept under the rug, along with other unpleasant memories of war and suffering.

At the time, it was called the Spanish flu, but we now know that it didn't originate in Spain. The science at the time wasn't much clearer about the causes. Although medical science had a phenomenological concept of influenza, the causative agent remained elusive. In a previous pandemic, the bacterial species *Haemophilus influenzae* was blamed for and named after influenza, but it turned out to be innocent.

Only in the 1930s was the underlying pathogen identified as a virus, and only in the 1990s was the connection to avian influenza established. Now we know that it was an H1N1 strain of influenza A virus that caused the 1918 pandemic. The strains are categorized by their specific types of haemagglutinin (H) and neuraminidase (N) genes. We now know to look out for new variants of bird flu that might acquire the ability to spread between humans. What caused the 1918 pandemic to become so deadly, however, has remained a mystery.

To find out more about the 1918 pandemic virus, Livia Patrono from the Robert Koch Institute in Berlin, Germany, and colleagues analyzed 13 formalin-fixed lung specimens dated between 1900 and 1931 from museum collections held in Berlin and Vienna. They identified influenza A virus in three of the specimens and produced one complete and two partial genome sequences of the virus.[3] Previously, only two complete sequences were known, both from victims in the United States, along with additional genetic data concerning the haemagglutinin gene. Suitably preserved samples from victims of the pandemic are very rare and hard to find.

Comparing the newly sequenced genomes to the known data from different places and times of the pandemic, Patrono and colleagues identified mutations that may be related to the virus adapting to the human host during the outbreak, which in turn may be linked to the observation that the second wave was deadlier than the first. The researchers were able to confirm that the H1N1 strains still circulating as seasonal flu today are descendants of the pandemic strain.

Considering the struggle to find samples relevant to the flu pandemic that occurred only a century ago, the research into medieval plague is flourishing remarkably. The notorious Black Death was the first wave of the second recognized plague pandemic. It swept across Europe from the year 1347 to 1352, killing up to half of the population in some areas. Subsequent waves kept coming until the early nineteenth century but caused fewer casualties.

The Black Death was long assumed to have emerged from central or East Asia, as its progress from the coasts of the Black Sea toward the Mediterranean ports is well documented in history books. One archaeological find that has been discussed as potentially relevant is a pair of burial sites near Lake Issyk Kul, in what is now Kyrgyzstan. Excavations in the nineteenth century revealed tombstones dating from 1338 to 1339 with inscriptions in the Syriac language citing a mysterious "pestilence" as the cause of death. Maria Spirou from the University of Tübingen, Germany, and an international team used modern technology to analyze the ancient DNA in the human remains from these burials and successfully identified the plague pathogen, *Yersinia pestis*, in graves dated to 1338.[4]

Genome analyses showed that this was not simply one of the many sites visited by the pandemic. The bacterial strain found matches the requirements for the common ancestor of the diversity that emerged during the Black Death, suggesting that the researchers found the source population, complete with the historical place and time that it ravaged communities in central Asia.

This remarkable discovery also enabled the researchers to investigate the likely animal source of the outbreak, as plague is known as a zoonotic disease that has permanent animal reservoirs and crosses into the human population occasionally.

"We found that modern strains most closely related to the ancient strain are today found in plague reservoirs around the Tian Shan mountains, so very close to where the ancient strain was found. This points to an origin of Black Death's ancestor in Central Asia," explained Johannes Krause from the Max Planck Institute for Evolutionary Anthropology in Leipzig, Germany, a senior author of the study.[5]

While the further progress of the Black Death across Europe is well documented, details of its impact can still be subject to debate. In a recent study, Adam Izdebski from the Max Planck Institute for the Science of Human History (now the Max Planck Institute of Geoanthropology) in Jena, Germany, and colleagues used a "big data paleo-ecology" approach to investigate its impact.[6] The fundamental idea is that in pre-industrial times, the land surface under cultivation was a direct function of the population available to work the fields. Thus, a population collapse of 50% or more should clearly lead to a reduction in crops being grown.

Studying pollen records from 261 radiocarbon-dated coring sites across Europe, Izdebski and colleagues found that the impact of the Black Death was as devastating as expected in many places but that some regions, including Ireland, Iberia, and parts of eastern Europe, appear to have had a lucky escape and carried on their agricultural business as normal. While specific reasons for the dramatic differences in impact remain unknown, the authors suggested that local cultural, demographic, economic, environmental, and societal contexts could have shaped the different outcomes.

Before the Black Death, many "plagues" hit early Eurasian civilizations, which may or may not have been caused by *Yersinia pestis*. The Justinianic Plague, which in 541–542 CE hindered Eastern Roman (Byzantine) emperor Justinian's attempts at restoring the empire's unity and glory, is currently the earliest plague pandemic recorded by history and confirmed by DNA evidence. *Y. pestis* DNA from that period was first detected in victims in Bavaria, later also in France, Spain, and England. The genomes revealed that this earlier pandemic strain is unrelated to the Black Death strains and both pandemics likely go back to separate zoonotic transfer events.

The discovery of multiple plague carriers in an Anglo-Saxon burial site at Edix Hill in Cambridgeshire, England, in particular, may require a change to current ideas about the spread of the pandemic, as Peter Sarris from the University of Cambridge, United Kingdom, argued.[7] Rather than spreading from Constantinople across Europe, the disease may have reached England on a separate route at the same time or even before it arrived at Justinian's capital.

Sarris also took issue with publications from Lee Mordechai and Merle Eisenberg from the National Socio-Environmental Synthesis Center in Annapolis, who argued that the Justinianic Plague was "inconsequential" in that it did not disrupt economic and political processes to a significant degree.[8] In a debate reminiscent of some of the arguments around the COVID-19 pandemic, Sarris highlighted that Justinian had to issue specific legislation to ward off the economic fallout of the plague. Importantly, Sarris concludes, the pandemic went far beyond Justinian's empire: "In both the Early Medieval Irish and Early Islamic worlds, for example, the era of the Justinianic Plague would be remembered as one of unprecedented mortality and pestilence."[9]

Further evidence of the impacts of the plague even before the first pandemic comes from Max Planck Institute for Evolutionary Anthropology researchers, including Maria Spyrou, Johannes Krause, and others, who described genomes of *Yersinia pestis* and *Salmonella enterica*, the pathogen causing typhoid fever, from the end of the early Minoan period in Crete.[10] The authors argued that, based on their discovery, infectious disease should be reconsidered as a possible agent of change in the transformations of early complex societies.

Going back even further in time, Aida Andrades Valtueña from the Max Planck Institute for Evolutionary Anthropology and colleagues sequenced *Yersinia pestis* genomes up to 5,000 years old from across Eurasia.[11] The researchers identified two separate strains of the pathogen that evolved in parallel and across a wide geographic area, with one being more flea-adapted than the other. The authors concluded that the spread of plague in that time was linked to increasing mobility and more widespread use of domesticated animals.

Meriam Guellil from the Estonian Biocentre in Tartu and colleagues discovered prehistoric roots of two other human pathogens, herpes simplex 1 and Hib (Haemophilus influenzae serotype b).[12] Both were present at the Edix Hill burial site of Anglo-Saxon plague victims in Cambridgeshire, England.

The impact of pandemics appears to be most dramatic when we can observe them derailing the machinery of a well-described civilization, hence the European bias in the examples discussed. Death and disruption will have been similar, if less well-documented, in the Americas when the Europeans brought smallpox and other diseases to which the Indigenous population had no immunity. The scale of the depopulation of the Amazon and the resulting ecological changes are also the subject of ongoing research and debate (see chapter 4).

To better understand the dangers and long-term perspectives of the COVID-19 pandemic and possible threats like mpox and new flu strains, it is most valuable to study the pandemics of the past, with their detailed historical context and increasing amounts of molecular genetic details becoming available.

The 1918 flu pandemic has already provided warnings relevant to COVID-19, such as the effects of super-spreader events. The way the H1N1 strain subsequently turned into seasonal flu also informs our thinking of what may happen to the pandemic coronavirus in the long term.

The three plague pandemics provide a warning regarding the longevity of human diseases. The first two stretched out over several centuries, and the third, which started in the mid-nineteenth century and spread *Y. pestis* around the globe, was only cut short thanks to antibiotics. If a multi-drug-resistant plague pathogen arises, we may find ourselves back on square one.

All these warnings from history are useful only to the extent that people are prepared to heed them. If COVID-19 is forgotten as fast as the 1918 flu was and the anti-science mood of trivializing such diseases wins the day, this century—with a larger and more connected human population than ever—may well become the era of even deadlier pandemics.

How viruses cross species boundaries

The COVID-19 pandemic has shown vulnerabilities in many aspects of the globalized world we live in. The early suggestion that the pathogen may have originated in or at least passed through the endangered pangolin put animal trafficking in the spotlight. The human-wildlife interface remains an important risk factor for further disease outbreaks.

Like other virus infections that have caused concern in the last two decades, COVID-19 crossed the species barrier from an animal reservoir and demonstrated the ability to spread between humans, making it a further example of a zoonosis. Many zoonotic diseases that have established themselves in the human population, such as today's childhood infections like chicken pox, have evolved to become relatively harmless, enabling them to spread without endangering the survival of their host population.

This relatively peaceful co-evolution between some zoonotic pathogens and human populations is thought to have originated when agriculture enabled people to live in larger communities, creating effective population sizes big enough to sustain a permanent equilibrium with a weakened pathogen. By contrast, viruses that have made the transition more recently may still be more lethal. Thus, Ebola virus disease, which occasionally causes outbreaks in tropical Africa, where its natural reservoir is in bats, typically causes a very high rate of mortality among those infected and therefore must be confined at all costs (we shall return to Ebola later in this chapter).

A few outbreaks of severe respiratory disease caused by coronaviruses have caused concern since the beginning of this century.

They came from different animal sources, and the course of the epidemics have highlighted problems in the response systems.

Coronaviruses, a group of spherical RNA viruses with an envelope that in the electron microscope can appear like a star's corona, were discovered in the 1960s in the context of human respiratory diseases like the common cold. They were also found to cause respiratory disease in some animal species and diarrhea in others.

The first major outbreak of a new and dangerous zoonotic disease linked to a coronavirus was the epidemic of severe acute respiratory syndrome (SARS) in 2002–2003. Starting in November 2002 in southern China, the disease affected more than 8,000 people, with a mortality rate of just under 10%. After spreading to 17 countries, the outbreak was contained, and no cases of SARS have been reported since 2004.

Initial studies into the source of the SARS version of the coronavirus identified the Asian palm civet (*Paradoxurus hermaphroditus*) as a carrier. This small mammal of the tropical forests was on sale for its meat at the local meat markets in Yunnan Province, China.

Further studies indicated that the civets may have been infected by bats. In 2017, the teams of Zheng-Li Shi and Jie Cui at the Wuhan Institute of Virology, China, discovered a mixed population of several species of horseshoe bats in a cave in the Yunnan Province carrying a coronavirus variant with precisely the same genetic elements that were found in the pathogen of the SARS outbreak.[13]

Bats are known to have a remarkable tolerance of viruses that may cause serious disease in other species. This feature has been linked to the dampened inflammation response to stress factors, such as DNA present in the cytosol, which can be a sign of viral infection but in the case of bats can also be linked to the stress caused by the high amount of energy needed for powered flight (more details further on in this chapter). Keeping the inflammation response under control enables the bats to mount a robust immune reaction against the virus while avoiding the side effects, such as swelling and fever.

Cara Brook from the University of California, Berkeley, and colleagues analyzed and modeled these effects in detail using cell cultures derived from two different species of bats.[14] The authors concluded that viruses respond to robust bat immunity by establishing more rapid cell-to-cell transmission rates than comparable viruses use in other groups of mammals. Therefore, when they transfer to humans, the viruses may become more deadly than established human viruses.

People don't often mingle with bats, but Brook and colleagues noted that the diseases are often transferred from bats to humans via another mammalian host, like the palm civet in the case of SARS. Moreover, they pointed out that disturbing bat habitat may stress the bats and makes them shed more viruses via their saliva and excretions. Working with the bat monitoring project, BatOne-Health, operating in Madagascar, Bangladesh, Ghana, and Australia, Brook and colleagues have been exploring the link between the loss of bat habitat and the transfer of their viruses into other animals and humans.

The next significant outbreak of coronavirus disease was the Middle East respiratory syndrome (MERS), first identified in Saudi Arabia in 2012. Like SARS, this disease is caused by a virus of the genus *Betacoronavirus*, but it is of a clearly distinct lineage. Compared to SARS, it is more deadly but less infectious in the transmission between humans. According to World Health Organization (WHO) figures from January 2020, there had been just over 2,500 cases, causing 862 deaths. Despite the high mortality of 34%, this virus is not regarded as a major global health threat, as its spread can be contained relatively well.

Virus strains identical to the one in MERS patients were discovered in dromedary camels. The virus appears to have passed
from camels to humans on several occasions, causing further outbreaks in 2015 and 2018. Following the first outbreak in 2012, government advice to avoid close contact with camels led to a backlash
among farmers in Saudi Arabia. The BBC reported in 2014 that
farmers kissed their camels in defiance of health advice. This kind
of resistance to medical advice might explain the later recurrence
of the disease. As recently as November 2019, several new cases of
MERS were reported in Saudi Arabia. All patients were known to
have had contact with camels but not with other people infected
with the virus.

Sohail Hassan from the University of Veterinary and Animal
Sciences in Lahore, Pakistan, and colleagues tested for MERS antibodies in camel handlers and their families in Pakistan, where
no cases of MERS infection have been reported.[15] Of 91 ELISA
tests (enzyme-linked immunosorbent assay, a standard immunological method), more than half were positive for MERS-CoV antibodies. With a 50% reduction plaque-reduction neutralization test
($PRNT_{50}$), 12 participants tested positive for exposure to the coronavirus, and 10 of these were confirmed by immunofluorescence.
While the authors interpreted the high ELISA result as a possible
cross-reaction from other coronaviruses, they considered the 10
positive tests obtained by the other two methods as an indication
that exposure to MERS-like coronaviruses is widespread among
camel handlers in Pakistan. A separate study conducted in Sudan
detected MERS-CoV antibodies in nearly all the dromedary camels tested but not in humans exposed to them.[16]

As transmission between humans mainly occurs within healthcare environments, Emmie de Wit and colleagues at the Laboratory of Virology of the National Institute of Allergy and Infectious

Diseases in Hamilton, Montana, tested the prophylactic use of the antiviral remdesivir.[17] In a study using rhesus macaques, the researchers found that the compound, if given before infection with the virus, protected the animals from becoming ill.

The coronavirus responsible for the COVID-19 pandemic is less deadly than MERS and SARS but also more infectious. Even with a death rate below 1%, it has still managed to kill millions.

The outbreak emanated from the city of Wuhan, the capital of the Hubei Province in China, and has been traced back to the Huanan Seafood Market, where numerous animals, including fish, chickens, pheasants, and wild animals such as bats, marmots, venomous snakes, and deer, are on sale.

In the early days of the pandemic, it appeared certain that the outbreak, like SARS and MERS, was a new zoonotic transfer, but the animal source remained an issue of debate. Sequencing of virus genomes from five of the first cases identified established that the same virus was present in all five patients, that it was new in that it had limited similarity to the SARS virus (79%) and to the MERS one (52%), and that it may have originated in bats, as it showed higher similarity (87%) to two viruses found in Chinese horseshoe bats.[18]

The connection between horseshoe bats and humans may again have passed through another carrier species. A new zoonotic connection emerged in the shape of the endangered Asian pangolin.

Just before the pandemic, in October 2019, Ping Liu, Wu Chen, and Jin-Ping Chen from the Guangdong Institute of Applied Biological Resources in Guangzhou, China, reported that they had discovered high loads of coronavirus in blood samples of 21 Malayan pangolins (*Manis javanica*) that had been confiscated from wildlife traffickers.[19] Many of these animals were ill, and 16 of them died despite rescue efforts. Although the motivation of the study was rooted in conservation concerns, as the Malayan pangolin is critically endangered, the discovery of the viruses led the researchers to this conclusion: "Malayan pangolins could be another host with the potential of transmitting the SARS coronavirus to humans. As a consequence, the viral metagenomic study of Malayan pangolin is meaningful both for the conservation of rare wild animals and public health."

Researchers Yongyi Shen and Lihua Xiao from the South China Agricultural University in Guangzhou compared the coronavirus sequences from the pangolins to those of the new outbreak and found them virtually identical.[20] The authors concluded that the pangolin may have had a role as an intermediate host between a bat reservoir and the human outbreak, in analogy to the role that palm civets played in the SARS epidemic. They also warned that transmission via pangolins could pose a threat of future outbreaks and that the animals should therefore be monitored for viruses.

These findings cast a spotlight on one of the most endangered and most trafficked groups of mammals. Pangolins, which are armored with tough scales like armadillos, have the defense strategy of curling up into a ball like a hedgehog. Unfortunately, this makes them vulnerable to human hunters, who can just pick them up and take them home. Hunted both for their meat and for their keratin scales (which are used in traditional medicine), the four species of Asian pangolins are all endangered, three of them critically. The four African species (two of genus *Phataginus* and two of *Smutsia*) are listed as vulnerable or endangered.

The critically endangered Asian pangolin species *Manis pentadactyla* and *Manis javanica* are featured on the EDGE list of unique and endangered mammals (see chapter 5). The Zoological Society of London, United Kingdom, which compiled this ranking, has championed the case of pangolin conservation, which until 2012 didn't find much attention.

One might hope that the news of pangolins carrying coronaviruses could help deter the poachers and traffickers, but then again, the reports of farmers kissing their potentially virus-infested camels demonstrated very clearly that rational arguments don't always win at the human-animal interface.

Yet another possible source of the COVID-19 pandemic emerged in March 2023 in the shape of the common raccoon dog (*Nyctereutes procyonoides*), which in China is bred and traded for its fur. Florence Débarre, a researcher at the French National Centre for Scientific Research, discovered sequences from COVID-19-positive samples taken at the infamous Huanan market in Wuhan, China, which turned out to represent raccoon dog DNA.[21] This could mean that this species, which was traded at the market, could have been the source of the outbreak that became the pandemic.

As researchers have warned repeatedly, there are many more virus strains out in nature that could jump the species barrier and cause further outbreaks. Disturbing habitats, hunting wild animals (whether it happens for food or medicine, as pets or as trophies), living too closely with domesticated animals (from camels to poultry)— all these human actions could lead to further outbreaks, which, due to the growing interconnectedness of a growing world population, will be even harder to control than the previous outbreaks.

To cause a global pandemic, a virus must cross many boundaries, from those between species to those between countries. Respecting both wild and domesticated animals and rethinking travel may help to stop future pandemics before they go global.

Preparing for the next Ebola epidemic

The Ebola virus disease outbreak that swept Guinea, Liberia, and Sierra Leone in 2014 reminded the world of the dangers of diseases transferred from animal reservoirs, or zoonoses (see above). They can be deadlier than diseases adapted to the human host, and once they establish a transmission chain between humans, it can become very hard to stop them. The West African Ebola epidemic became the largest of its kind by May 2014, and two years later the WHO estimated the total death toll to be 11,310, out of 28,616 reported cases.

As the epidemic unfolded, symptom treatment, such as rehydration therapy, helped to reduce the mortality rate, initially reported to be above 70% to below 50%, while systematic control of transmission routes, including infection during patient care and at funerals, eventually enabled the countries affected to end the outbreak. Since then, several smaller outbreaks have been reported, the largest of which killed 2,280 people in the Democratic Republic of the Congo and Uganda between August 2018 and June 2020.

These episodes illustrate that the disease remains a deadly threat, as it is poorly understood, there is no cure, and vaccines are still in experimental stages, so they can only be used as a last resort in situations of acute danger. Research is slowly catching up with the disease, which may come back to haunt us at any time.

As Ebola was quite rare prior to the 2014 epidemic and any work involving the live virus requires the highest security level, scientific knowledge about it still has some significant gaps, right down to the molecular level. Moreover, there are only a few closely related viruses to draw comparisons to. The genus *Ebolavirus* comprises five known species, four of which have caused Ebola virus disease in humans. Their closest relative is the Marburg virus (genus *Marburgvirus*), which has only caused a few outbreaks and is mainly of concern as a potential agent in bioweapons. Both genera belong to the family of filoviruses (Filoviridae), which also includes the genus of *Cuevavirus*.

The structure and function of Ebola virus is still incompletely understood. A better understanding of the mechanistic details of these viruses could yet lead to unexpected ways of treating the disease.

Until now, there is no known treatment for any disease caused by filoviruses, apart from the straightforward and sometimes life-saving alleviation of dangerous symptoms such as dehydration. Only a few leads toward possible treatments have emerged so far. As clinical trials for efficiency are impossible with such dangerous and normally rare diseases, a few potential treatments were tested during the West African epidemic based on animal results alone. None of them demonstrated unequivocal treatment success in the limited time while infected patients were available.

Blood transfusions from Ebola survivors have in some cases been used successfully as treatments and are backed by the WHO, which has issued preliminary guidelines for their use.

Jakob Nilsson from the University of Copenhagen, Denmark, together with Stephan Becker's group in Marburg, identified a new host factor necessary for the spread of an Ebola virus infection. Working in cultures of human cells, Nilsson and colleagues demonstrated that Ebola propagation recruits the human phosphatase enzyme PP2A-B56 to activate its own VP30 protein, the unphosphorylated version of which is necessary for transcription (the synthesis of RNA based on a DNA template).[22] Using a specific inhibitor of this phosphatase, the researchers could inhibit the transcription and spread of the virus, reducing it 10-fold.

A few vaccine candidates existed at the time of the West African epidemic. The candidate VSV-EBOV, based on the vesicular stomatitis virus in combination with an Ebola-typical glycoprotein, was used experimentally in Guinea during the epidemic, and a study published at the end of 2016 reported it to be highly effective in protecting against the disease.[23]

Researchers followed a carefully designed ring vaccination strategy, where more than 4,000 contacts (as well as contacts of contacts) of patients confirmed to be infected with Ebola were identified and considered for vaccination. Those who were eligible and gave consent were randomly assigned to either immediate vaccination or vaccination after a delay of 21 days. None of the contacts vaccinated immediately contracted the disease, while in the delayed control group of around 2,000 people, 23 cases of Ebola arose.

After advice from an independent safety board, randomization was abandoned and immediate vaccination offered to all eligible candidates identified in the contact rings around known cases. Overall, 5,837 individuals were vaccinated, of whom none was diagnosed with Ebola more than 10 days after immunization. Three patients suffered serious side effects, but all recovered without lasting harm. Thus, VSV-EBOV can count as the first efficient vaccine against Ebola virus disease. It was also used in the containment of the 2018 outbreak in the Democratic Republic of the Congo. Under the name of ERVEBO, the vaccine was granted a conditional marketing authorization by the European Commission and was approved by the WHO.

Immunizations are a highly efficient tool to rein in and even eradicate a known pathogen that is established in the human population. However, as the repeated worries around flu vaccines and the threats of new variants of avian influenza have shown, they have their limitations when dealing with zoonoses, as one never knows which variant of a pathogen might jump across the species barrier next. With COVID-19, humanity had a very lucky escape due to the vaccine being developed in record time. To limit the dangers of future zoonotic diseases, a better understanding of the ecological context is necessary.

While the specific animal source of Ebola outbreaks often remains unknown, there are known natural reservoirs and transmission paths. Fruit bats (family Pteropodidae), in particular, are known carriers of the virus, although they don't appear to get any kind of disease symptoms from it. Various primate species also get the disease, and in some cases, humans became infected from contact with ill or dead apes or monkeys.

Risks of infection with Ebola and other zoonoses from wildlife in tropical forests are linked with hunting and the bushmeat trade, which could be controlled more efficiently. Research suggests, however, that deforestation also increases the risks. Julia Fa from Manchester Metropolitan University, United Kingdom, in association with the Center for International Forestry Research, Indonesia, demonstrated a statistical correlation between Ebola outbreaks and recent deforestation.[24]

Studying satellite records of vegetation changes near 27 areas where Ebola outbreaks happened compared to 280 areas where there were none, the researchers found that the loss of dense forest cover was an important risk factor in Ebola outbreaks. While previous studies had hinted at such a connection, the new work adds a time dimension, suggesting that the risk is particularly high within two years of the deforestation.

The most plausible underlying reason would be that disease-carrying animals losing their habitat are more likely to encounter humans than those thriving in remote, untouched rainforests. It is also quite possible that the remaining areas of untouched tropical forests may harbor further strains of Ebola or other zoonoses that could cause even more damage than the ones that have emerged so far.

Additional lessons in resilience may be available from human survivors of the epidemics. The West African Ebola epidemic produced more victims than all previous outbreaks combined. With a mortality under 50%, it also produced more Ebola survivors than existed before. There are important questions to be asked regarding how long the virus can remain present in certain bodily fluids.

The large number of survivors will help to answer these questions more thoroughly than was possible before.

The WHO set up preliminary guidance specifying that male survivors should have their semen tested monthly and strictly observe safe-sex practices until they have tested negative at least twice. This is because the virus is known to persist in immune-privileged tissues, including the testicles. Persistence of viruses in reproductive tissues of female survivors is also a concern, especially if they were pregnant or breastfeeding when exposed to the virus.

The good news from studies of Ebola survivors is that immunity against the virus appears to be long lasting, sometimes even lifelong. The first documented Ebola epidemic, in 1976 the Democratic Republic of the Congo (then known as Zaire), left only 38 survivors. Eleven years after the event, some of the survivors were tested and found to still have an efficient immune response against the virus. A more recent study re-examined six survivors who had suffered a serologically confirmed Ebola infection in 1976, as well as eight others who had had the characteristic symptoms but no definitive confirmation. Using three independent measures of immune responsiveness to Ebola, Anne Rimoin from the University of California, Los Angeles, and colleagues found that serum antibodies from some of the survivors still responded positively to all three measures and were still fully able to neutralize live viruses.[25]

This finding confirms the value of the blood-transfusion treatment that was used in some cases during the West African epidemic. It also raises hopes that immunological treatment options could be developed based on the long-term immunity observed in survivors.

All in all, many valuable lessons have been learned from the West African epidemic, ranging from the challenges of quarantine procedures and efficient healthcare provision in such a crisis through to the value of ring vaccination strategies and other preventive measures. Education and awareness efforts have also been stepped up, as knowledge of transmission routes is crucial for the success of quarantine measures.

To better avoid further surprise epidemics of zoonotic diseases, however, an improved understanding of the disease mechanisms and their ecological background in wildlife will be necessary. Research is only beginning to answer these important questions, and the problem is that, as more human habitations encroach on dwindling wildlife habitat, the risks of zoonotic transmission are likely to grow.

CHAPTER 12

The Anthropocene and Beyond

The Bible tells us to "subdue the Earth," but it is now becoming apparent that our civilization has followed this demand a bit too enthusiastically, destroying ecosystems and disrupting Earth systems in the process. Early civilizations subdued and shaped their regional environments, but after the Industrial Revolution, and especially in the twentieth century, the changes became global. Many of them, like species extinctions, are irreversible. Others, like climate change, trigger a cascade of problems that we still can't fully comprehend, as we saw in chapter 10. We're changing the world, but not in a good way.

Human impact on Earth systems can be seen from space and will still be obvious in the geological record long after our species is extinct. Therefore, geologists have started the process of defining a new unit of geological time, the Anthropocene. They need to agree on a start time and on a geological marker that will still be obvious in a million years. Isotopes from nuclear weapons tests and plastic pollution have been suggested, as will be discussed in this chapter. Like several other breaks in the geological calendar, the Anthropocene looks set to see a mass extinction linked to its name (see chapter 5).

Although the official process isn't completed yet, the Anthropocene as a concept is already widely used and has proven fruitful for scientific discussions of the global change happening. It also offers the opportunity to look to a time after the Anthropocene and ask how Earth systems and biodiversity will recover after we're gone.

The world's vanishing lakes

Maps of the world are changing all the time, as countries split, reunite, or disintegrate. Within the last few decades, the Soviet Union, Yugoslavia, Czechoslovakia, and Sudan have fragmented, while countries and cities have also been renamed. Printers of maps and makers of globes not only have electronic competition to fear but also fast-moving political change.

One could hope that turning the switch of an illuminated globe to make it display the physical geography rather than the political boundaries would provide a more immutable face of the Earth. But on this level, too, our planet is changing.

Turning eastward from the Mediterranean, for instance, there is first the Black Sea, then the Caspian Sea (almost as large but landlocked), and next to it the Aral Sea, another large lake with no known outlet. In the 1960s, the Aral Sea was the fourth-largest lake on our planet, which is how it appears on the globe I have had since childhood. It supported a large fisheries industry and coastal biotopes such as the wetlands of the deltas of its major feeding rivers, the Amu Darya from the south (in Uzbekistan) and the Syr Darya from the east (in Kazakhstan).

In an attempt to convert central Asia's deserts into cotton plantations, Soviet government programs established irrigation systems that diverted large parts of the flow of both rivers. As a result, the water level in the Aral Sea dropped by around 20 centimeters (8 in.) per year in the 1970s and accelerated in later years as water use increased and very little was done to rein in what observers described as catastrophic loss of water through leaks and ineffective infrastructure.

After 1991, newly independent Uzbekistan continued the wasteful water use under the hard authoritarian regime of Islam Karimov, who remained president from independence until his death in September 2016. By 2007, the Aral Sea had shrunk to 10% of its former surface area, splitting into four separate basins. This version now appears in Google Maps, although the southern parts are still threatened by further decline. In 2014, for instance, the southwestern basin dried up entirely for a time.

Kazakhstan, meanwhile, built a dam in 2005 to control flow from north to south, which helped to save the small northern part of the lake at the expense of the rest. By now the exposed land has turned into an ecological disaster area. Heavily polluted salt residue left behind after evaporation gets dispersed by winds and blown onto the fields, which in turn will require even more irrigation to wash out pollutants. The absence of a climate-regulating body of water has already led to markedly more pronounced weather extremes. NASA has compiled satellite imagery demonstrating the dramatic changes in the area after the demise of the Soviet Union.

The disappearance of the Aral Sea may be one of the larger environmental disasters that humans have caused, but dozens of smaller variants are found around the world. International fame doesn't protect bodies of water, as widely known ones—including the Great Salt Lake (Utah), the Dead Sea (Israel/Jordan), Lake Chad (West Africa), and Urmia Lake (Iran)—are threatened, along with a myriad of less famous, smaller lakes. If we want to protect these water resources and the services they provide, it would be helpful to know more about them.

Counting and measuring lakes is notoriously difficult, not only because they may disappear or split up over time but also because it depends on criteria. The Caspian Sea (Eurasia) could be described as the world's largest lake, but then again, it could also be called an ocean, and it only lost contact with the world oceans some 11 million years ago. Lake Superior (United States/Canada) may be the largest freshwater lake, unless you count neighboring Lake Huron and Lake Michigan combined (as they are linked thus share the same water level). Lake Vostok is huge but hidden under the ice of the Antarctic.

Satellite maps show a myriad of smaller lakes, but geographers know surprisingly little about them. Estimates of fundamental parameters, like total surface area and total volume of lakes above a certain size, vary widely. Ecologically relevant data, such as the residence time of water in lakes and the release of greenhouse gases from their surface, are similarly insufficient.

Bernhard Lehner and colleagues at McGill University in Montreal, Canada, addressed this problem by newly estimating the volume and residency time for all lakes larger than 10 hectares (25 acres) in surface area and making these parameters available in a new database, HydroLAKES.[1] This now includes shoreline polygons of 1.43 million lakes and reservoirs.[2] More than half of these, around 900,000, happen to be in Canada, a wet legacy of the glaciers retreating at the end of the last ice age.

The database includes reservoirs, but the topography of reservoirs can typically be derived from maps made before the land was flooded, and they present different problems (e.g., the water level is controlled by humans and they are not natural ecosystems), so they are excluded from the following figures. As for surface area, the 1.42 million natural lakes larger than 10 hectares (25 acres) cover 2.67 million square kilometers (1.03 million mi.[2]) between them, slightly more than the Mediterranean Sea. Casting the net a bit wider, there are 21.2 million natural lakes larger than a hectare (2.5 acres), with 3.23 million square kilometers (1.25 million mi.²) of surface area, surpassing the area of the Mediterranean.

As with the oceans, detailed investigation of the topography hidden under deep water (bathymetry) is a massive and challenging task that is so far only completed in small parts. For the bulk of the smaller lakes, there is little chance of obtaining high-quality bathymetry data. Therefore, the McGill team developed a way of predicting the topography under the water based on the observable topography around the shores. Essentially, lakes set in rough, mountainous areas with steep slopes are likely to be deeper than those in flat areas. The origin of the topographic structure must also be considered. Lakes formed in volcanic or meteorite craters, for instance, may have steep slopes at the sides but a gently curved floor underneath most of their area.

The researchers trained their algorithm on lakes with known bathymetry data and then applied it to the complete database. Based on this method, they arrived at an estimate of just over 180,000 cubic kilometers (43,000 mi.³) for the total volume of these 1.42 million natural lakes larger than 10 hectares (25 acres). While this is three orders of magnitude less than the volume of the South Atlantic, the relatively large surface and long shorelines of the lakes mean that the lakes still represent an immense amount of aquatic and coastal habitats. The shorelines of the lakes studied added up to seven million kilometers (4.3 million mi.), which is four times longer than the ocean shores measured at a similar resolution.

For environmental and ecological considerations, it is important to know what goes into and what comes out of lakes. A key parameter is the hydrological residence time (i.e., the amount of time any given unit of water might be expected to spend in the lake before it flows onward or evaporates).

Lehner and colleagues have, for the first time, estimated a residence time for each of the lakes with more than 10 hectares (25 acres) of surface area. They estimate the global mean to be five years, with larger lakes cycling more slowly than smaller ones. Having at least reasonable estimates for such parameters is important when it comes to assessing the likely impact of human activities, from the deviation of water courses through to pollution.

The accumulation of pollutants is particularly problematic in endorheic lakes (those without an outlet) like the Aral Sea, particularly if they are shrinking and pollutants become more concentrated. Large parts of central Asia, as well as smaller areas of other continents, drain into such lakes, and the example of the Aral Sea shows that this situation can quickly become a recipe for environmental disaster if water resources are poorly managed.

But even under more favorable circumstances, lakes, like rivers, must cope with large amounts of pollutants that could put their ecosystem function at risk. The ongoing one-way street of phosphate use in agriculture is one of the biggest problems affecting many freshwater lakes and rivers. In many lakes, phosphate has accumulated to an extent that severely disrupts the ecosystem function by eutrophication.

As the problem is well known and a global introduction of phosphorus recycling is not in sight, many countries have turned to localized geo-engineering experiments trying to redress the balance. In an editorial for a special issue of the journal *Water Research*, guest editor Miquel Lürling from the University of Wageningen in the Netherlands and colleagues compiled more than 100 different approaches from around the world and evaluated them.[3]

Most treatments used chemicals such as aluminum salts or lanthanum-modified bentonite additions to control phosphorus. However, as co-editor Bryan Spears from the Centre for Ecology & Hydrology, United Kingdom, cautioned, the outcomes observed were very variable. Thus, he noted, careful small-scale trials are needed to assess what kind of treatment might work in each environment. Moreover, he warned that such treatments should not be seen as a cure for continuing bad environmental practice or a replacement for improvements.

Plastic waste, which we will come back to later in this chapter, is another pollutant that is increasingly being felt in lake environments, in addition to its accumulation in the oceans and its impact on marine ecosystems. Applying methodologies already established in the marine environment, Matthew Hoffman and Eric Hittinger from the Rochester Institute of Technology, New York, estimated the total amount of plastic that enters the Great Lakes and simulated its distribution with flow models.[4]

The researchers concluded that almost 10,000 metric tons (11,000 tons) of plastic enter the Great Lakes every year; half ends up in Lake Michigan and a quarter in Lake Erie. For comparison, earlier estimates of plastic pollution entering the oceans from the entire length of US coasts have ranged from 40,000 to 110,000 metric tons (44,000 to 121,000 tons).

In the oceans, much of the plastic accumulates in the five large gyres. Lakes have a much smaller volume in relation to the shorelines where the pollution is produced and released, and thus they have no space to hide the plastic. Prevailing winds and currents may carry the pollutants from leading sources, including major cities such as Cleveland, Detroit, Chicago, and Toronto, to other shores or islands where it accumulates. Modeling these flow patterns should help with cleanup operations, but it doesn't remove the fact that our civilization is still releasing too much plastic waste into the environment, which puts vital functions of freshwater and marine ecosystems at risk.

Lakes around the world provide habitat and ecosystem services far beyond their obvious roles for water provision, fisheries, and recreation. The larger ones, acting like the oceans, serve climate regulation and the hydrological cycle. The Caspian Sea, for instance, takes up and recycles the flow of Europe's longest and largest river, the Volga. As noted above, the loss of these functions is already being felt around the area of the former Aral Sea.

Other important roles rely on the extensive shorelines and adjacent shallow waters. Throughout much of Africa, for instance, the papyrus plant (*Cyperus papyrus*) grows in sunlit swamps and shallow lake waters. Ancient Egyptians used it for a variety of purposes, from making papyrus paper to building ships. Even while it is still in the water, the plant helps to keep the lake ecosystem clean and aerated.

Today, the livelihoods of millions of Africans depend on these lakeshore ecosystems, as research by Anne van Dam and colleagues at the UNESCO IHE Institute for Water Education in the Netherlands has established.[5] These ecosystems and livelihoods are under threat from population growth and land-use change. The challenge for researchers and local communities is to establish a sustainable modus of utilizing the papyrus wetlands without destroying them.

Some restoration projects are already underway. In 2006, researchers led by David Harper from the University of Leicester, United Kingdom, raised the alarm that the ecologically important environment of Lake Naivasha in Kenya was under threat, as the water level receded and the papyrus lining the shores was degrading due to population pressures.[6]

Working with the flower-growing companies of the area, which produce a significant part of Kenya's foreign income, Harper and colleagues developed and promoted sustainable management strategies. For instance, in an innovative approach introduced in 2012, Harper and colleagues released floating islands to grow papyrus on. The islands were produced by a US company using post-consumer plastic waste, such as water containers. Unlike the coastal papyrus under threat from an expanding buffalo population, these floating islands are better protected and can thus more reliably continue to provide filtering and aeration for the lake water. Moreover, they were anchored deliberately at the mouth of the Malewa River, which feeds the lake, in the hope that the papyrus will stop the lake from silting. As well, the roots of papyrus islands act as important fish nurseries and feeding grounds, while their five stems, growing up to five meters (16 ft.) tall, hold a rich biodiversity of birds, such as warblers and kingfishers.

The project was supported by the German supermarket chain REWE, among others, which only sells roses with the fair-trade label and thus is exposed to scrutiny into its production methods in Kenya. According to Harper, since 2010, Lake Naivasha and other lakes in the area have recovered due to rising water levels, helped by more sustainable water management.[7]

Around the world, wherever water resources are degraded due to poor management or ineffectual response to changing parameters, there is the potential not only of severe damage to ecosystems but also of economic loss and social unrest. In Iran, the rapid shrinking of one of the countries natural wonders, the high salinity Lake Urmia, is likely to affect tourism income. The lake is nominally protected as a national park, but it has lost 90% of its surface area due to damming of rivers and heavy water use in its basin. The receding water level and increase in salinity poses problems for many migratory birds, such as flamingos, ibises, spoonbills, and storks, that have so far used the many islands in the lake as stopover points but may be set to lose this opportunity. In recent years, fears that Lake Urmia may dry out entirely have sparked protests from environmentalists and some government activity.

In the highly unstable region of West Africa between the countries Nigeria, Niger, Chad, and Cameroon, the shrinking of Lake

Chad and the desertification of its shores has diminished the livelihoods of local communities and reportedly contributed to the
rapid rise of the anti-Western terror militia Boko Haram, showing
that humans alter physical geography at their peril.

Adding new lakes to the landscape in the shape of reservoirs for
hydroelectricity, long seen as a green solution, has recently been
recognized as another possible source of environmental problems.
As many countries, such as China and Brazil, invest heavily in hydroelectricity, research has shown that reservoirs built in unsuitable locations, especially in tropical climates, may emit greenhouse
gases on a similar scale as the fuel use they replace.

The map of global water resources is changing too rapidly for
scientists to be able to assess the situation and launch rescue efforts. Linking global and local efforts to avert multiple crises in
water security and lake ecosystems is now a highly urgent task.

Losing the soils that feed us

The year 2015 was declared the International Year of Soils by the
United Nations General Assembly. The United Nations Food and
Agricultural Organization (FAO) campaigned to raise awareness
globally, organizing events in many different countries and cele-
brating small indications of progress, such as the fact that soils are
now mentioned in the UN's Sustainable Development Goals, which
were adopted in 2015 and replaced the Millennium Development
Goals, although soils are not headlining one of the 17 main goals.

Some model projects were launched in developing countries
and well-meaning coverage in media and scientific journals duly
followed. However, a major U-turn in the way we look after the
ground beneath our feet failed to materialize. A drastic change of
course is urgently required, as the report *Status of the World's Soil*

Resources demonstrates, which the FAO released at the end of the Year of Soils.[8]

The over-600-page report, prepared by FAO's Intergovernmental Technical Panel on Soils, is based on the work of around 200 soil scientists from 60 countries. Its overall conclusion is that more than half of the world's soil resources are in a less than good condition, ranked as fair, poor, or very poor. A third of the land was described as moderately or highly degraded.

The 10 main threats to soil quality covered in the report are soil erosion, soil organic carbon loss, nutrient imbalance, soil acidification, soil contamination, waterlogging, soil compaction, soil sealing, salinization, and loss of soil biodiversity.

Of these, erosion accelerated by human activities is possibly the most urgent problem, as it threatens to remove much of the existing agriculturally valuable topsoil from the land surface within half a century. Typically, erosion speeds up dramatically when natural vegetation is cleared and replaced with crops. This happens because the complex natural system of many interdependent species is replaced with a single crop species, which will not cover the entire surface and may not provide sufficient root density to stabilize the ground. In addition, land left bare after harvest or when it is not used for economic reasons is particularly exposed to the risk of erosion. The report estimates that erosion removes 25–40 billion metric tons (27.6–4.4 billion tons) of topsoil per year, leading to cereal production losses of 7.6 million metric tons (8.4 million tons) per year. Projections indicate that the global loss of soil by 2050 will be equivalent to the entire arable land surface of India. Unless trends are reversed, the world's soils could be gone within 60 years.

Unsustainable agricultural techniques are a large part of the problem. Agricultural intensification in Europe in the twentieth century has dramatically increased erosion. A study of sediments in a pond in France showed that sediment accumulation had increased 60-fold over the last century, as the surrounding area converted to intensive agriculture.[9]

In some parts of Africa, such as in Madagascar, slash-and-burn agriculture has been established as a standard procedure; this method relies on clearing forests and then cultivating the land only for a short period of time, taking its loss to erosion as a given, rather than doing something to prevent it.

In Europe, the British biologist and writer George Monbiot has often criticized the European Union agricultural policy for handing out incentives for keeping land in farmable condition, even if its actual use would be uneconomical and letting the land rewild with native shrubs would offer environmental benefits and protection from erosion. Monbiot has argued that these inappropriate incentives fuel erosion and increase the flood risk.

Some progress has been made in the fight against erosion—for instance, by minimizing tillage (moving the soil; e.g., by plowing). Encouraging trends have been reported—for instance, for croplands in the United States and in some parts of Latin America. However, many other parts of the world still have unsustainably high levels of erosion and deteriorating trends.

An additional loss of agriculturally valuable surface is due to the accelerating urbanization of the world population and expansion of cities. Historically, settlements have tended to nucleate and grow to become towns and cities in places benefiting from fertile soil. With their expansion, they have often covered large areas of the very ground that fueled their growth.

Even when the soil stays in place and isn't covered up with asphalt and concrete, pollution and intensive agriculture may degrade its quality by disturbing its chemical balance. The massive use of synthetic fertilizers to provide nitrogen and phosphorus compounds to crops, for instance, has not only upset the global cycles of these elements but also changed the chemical composition of soils locally. Pollution, ranging from excess pesticides to toxic metals from industrial activities, also deteriorates the quality of the soil.

Salinity in the soil can affect plant growth. Unsustainable water use for irrigation purposes can lead to salt enrichment in water courses and can stress ecosystems located downstream. The 2015 FAO report concludes that large, poorly designed irrigation projects are the largest cause of human-made salinization of soils.

Moreover, where lake or ocean water evaporates, excess salt distributed by the wind can cause problems, such as in the area surrounding what used to be the Aral Sea (see above). Based on the 2015 report, the FAO concluded that "human-induced salinity affects an estimated 760,000 square kilometers [293,000 mi.^{2}] of land worldwide—an area larger than all the arable land in Brazil."[10] As a possible remedy, the FAO suggested there was "significant potential in Africa for the development of local-scale distributed irrigation systems that rely primarily on near-surface groundwater that is replenished annually."[11]

The pH of the soil is another important parameter that can be affected by human activities. There is a natural global division into areas of acidic and alkaline soils related to the balance of precipitation and evaporation. It has long been known that alkaline soils tend to occur in arid areas, while humid regions have acidic soils. Eric Slessarev from the University of California, Santa Barbara, and colleagues established in 2016 that this division applies globally and with a remarkably sharp threshold.[12]

The explanation behind the phenomenon is that soil is typically buffered by two chemicals whose concentrations respond differently to leaching. Calcium carbonate buffers in the alkaline range near a pH of 8.2. In arid climates, this mineral usually determines the pH of the soil. When precipitation exceeds evaporation, however, the highly soluble carbonates are washed out, leaving the more immobile aluminum compound $Al(OH)_3$ in control of pH, which buffers at a pH of 5.1.

With their analysis of more than 60,000 pH measurements from around the world, Slessarev and colleagues found that this simple principle can explain 42% of the global variation in soil pH. It can serve both to highlight areas where a deviation from the expected pH may point to chemical factors yet unrecognized and to predict soil chemistry where no measurements are available.

This natural two-state system regulated by rainfall can be severely disturbed by human activities, including intensive agriculture. Analyses in China reported by F. S. Zhang from the China Agricultural University in Beijing and colleagues showed that the pH of croplands declined significantly from the 1980s to the first decade of this century, as agriculture was intensified.[13] The authors found that chemical processes related to the cycling of the nitrogen fertilizers used accounted for much of the pH change observed, with direct application of acidic substances playing a less important role.

Agriculture has always tried to simplify the ecosystem it works with in order to grow the maximal amount of the desired crop or livestock species. Thus, modern pesticides and herbicides serve to reduce the number and biomass of other, less appreciated species.

In contrast, ecologists have increasingly revealed how all life, including crop species, depends on a complex network of interactions with other species. While a naïve bookkeeping approach to agriculture might suggest that a reduction in species number means more nutrients and thus more output for the desired crop species, ecology has often shown that natural complexity is a force for good.

The complexities are possibly greatest and most underappreciated underground. Unlike a hydroponic setup with inert substrate, water, and a limited number of chemicals, a natural soil contains many thousands of species in every handful of dirt. As we saw in chapter 1, these include invertebrates, fungi, bacteria, and even bacteriophages.

Science and agricultural technology have long neglected the importance of invertebrates, such as the humble earthworm, for soil quality. Likewise, the complexities of intra- and interspecies interactions and chemical communication around plants, both above and below ground, are only now beginning to be appreciated (see chapter 1). Two studies published in 2017 provide glimpses of how the interactions in the soil can determine the diversity of plant communities.

François P. Teste from the University of Western Australia in Crawley and colleagues studied plants that rely on five different nutrient-deriving techniques. They collected soil from the ground below these types of plants and then tried different combinations of these soil biota, as well as sterilized soil, on different species. With the data obtained from experiments with individual plants, the researchers then conducted simulations of how plant communities might develop under various soil conditions.[14]

The authors could establish a wide range of both positive and negative feedback effects, meaning that the soil influenced by a given tree species may promote or hinder the growth of saplings of the same species. It is to be expected that negative feedback, in hindering saplings trying to establish themselves near their parent tree, would enhance diversity. However, the simulations adding in interactions one by one showed that both positive and negative feedback effects can contribute to diversity.

Separately, Jonathan Bennett from the University of British Columbia Okanagan, Canada, and colleagues reported a field study of feedback interactions in soils and seeds from 550 populations of 55 North American tree species.[15] These trees use two groups of feeding strategies that are both based on fungi—namely, symbiosis with either ectomycorrhizal fungi (EM), which form a thick sheet around root tips, or arbuscular mycorrhizal fungi (AM), which penetrate into certain cells of the plant root.

Bennett and colleagues found that the soil around EM trees tends to further the growth of the tree's own seed, while the environment of AM trees tends to block it. Both responses can be understood as different strategies against the pathogens lurking in the soil. While EM fungi appear to be well suited to defend the saplings against the enemies of the parent plant already present in the soil, the plants with AM fungi follow a strategy that is known as the Janzen-Connell effect—namely, to establish their offspring out of the reach of their own pathogens, thereby increasing forest diversity, as Wim van der Putten explained in a perspective article accompanying the two papers.[16]

What is to be done to make our use of the remaining soils more sustainable? The 2015 FAO report identifies four priority areas: minimization of degradation and restoration of degraded soils, specifically where food security of vulnerable populations is at risk; stabilizing the current level of soil organic matter and organisms; stabilizing or reducing the global throughput of nitrogen and phosphorus fertilizers; and improving knowledge and awareness.

These will, however, remain idle wishes unless things start happening on the ground, as it were, and in the places where agricultural policies are set into law. There were several specific policy suggestions based on the report, including the FAO suggesting, "Introducing appropriate and effective regulation and incentives. This could include taxes that discourage harmful practices such as excessive use of fertilizer, herbicides and pesticides. Zoning systems can be used to protect the best agricultural soil from urban sprawl. Subsidies could be used to encourage people to purchase tools and other inputs that have a less harmful impact on soils, while certification of sustainable crop and livestock practices can make produce more commercially attractive at higher prices."[17]

Addressing taxes and incentives is particularly important because existing ones, including those under the European Union's common agricultural policy, perversely encourage practices that enhance soil erosion, as explained above. Even some supposedly environmental subsidies, such as those for biofuel production, ended up supporting vast, unsustainable monocultures that don't even meet the original goal of improving the carbon balance.

Alternative agricultural practices, including "no tillage" approaches, crop rotation, and permaculture, exist but are insufficiently known and promoted. Permaculture, for example, offers a complete philosophy based on human activities fitting in with the ecological context. The concept of "permanent agriculture" as a system that can be sustained indefinitely corresponds to what would be 100% sustainability.

Government subsidies currently propping up unsustainable land use could be redirected to become tools that enable farmers to switch to sustainable practices. Farmers in many parts of the world are suffering from intense competition for the lowest prices brought on by globalized food trade. Most will have no time or energy spare to reinvent agriculture or to think about the long-term consequences of their daily grind. Evidence-based policy taking into account the natural benefits of ecological complexity of our soils will have to drive substantial changes to practices worldwide if we are to produce food for 10 billion people toward the end of the century.

Our planet wrapped in plastic

Henderson Island is as far from civilization as one can get on planet Earth. There is no major human habitation or industrial activity within 5,000 kilometers (3,100 mi.) of the island, which is part of the United Kingdom's Pitcairn territory in the South Pacific and recognized as a UNESCO World Heritage site. The nearest human settlement is on Pitcairn Island, 115 kilometers (71 mi.) away, which has around 40 inhabitants (some of whom are descendants of the mutineers of HMS *Bounty*).

And yet, when Jennifer Lavers from the University of Tasmania, Australia, visited Henderson and combed its sandy beaches from May to August 2015, she found them covered in plastic waste. At a density of more than 670 items per square meter (per 10.8 ft.²), the waste level was the highest recorded anywhere outside an actual landfill site.[18] Lavers extrapolated that the island must contain at least 37 million items of debris weighing a total of 17.6 metric tons (19.4 tons).

A map of ocean currents yields an explanation of the phenomenon. Henderson Island is within the western reach of the South Pacific gyre, which rotates counterclockwise. So it sweeps up the Chilean coast, picking up all waste washing down from the western flank of the South American continent, and then dumps some of it on the Pitcairn Islands on its way back south.

The North Pacific gyre has long been recognized as a marine site of waste accumulation and is known colloquially as the Great Pacific Garbage Patch. With the globalization of plastic production and use, the North Pacific gyre's southern mirror image has become a similar carousel of plastic waste. Remote islands like Henderson simply act as a sink filtering some of the rubbish out of the oceans and allowing scientists a close-up assessment of the scale of the problem.

Roland Geyer from the University of California, Santa Barbara, Jenna Jambeck from the University of Georgia, and Kara Lavender Law, from the Sea Education Association, Massachusetts, reported the first comprehensive quantitative assessment of all plastic ever produced, including its chemical composition, its useful lifetime, and its fate after use, leading to predictions of plastic-waste accumulation. Previously, Geyer and colleagues had estimated plastic-waste quantities based on their proportion in the solid waste handled.[19]

The newer research, by contrast, started from production data and takes into account the different kinds of uses, ranging from instantly disposable packaging to plastic products that stay in use for years.[20] This results in waste streams that occur with predictable delay after production, enabling the authors to forecast plastic-waste generation for the coming years until 2050.

The researchers found that between 1950 and 2015, global production of plastics, including resins and synthetic fibers, increased from 2 million to 380 million metric tons (2.2 million to 419 million tons), with the cumulative production totaling 7.8 billion metric tons (8.6 billion tons)—more than enough to wrap the entire planet in a layer of plastic wrap. The exponential growth curve corresponds to an annual rate of 8.4%, or more than twice the rate of the global economic growth.

The steep exponential growth also means that a large part of the total was produced relatively recently. The authors estimate that just under a third of the plastic produced until 2015 was still in use at the end of the period studied. The rest has ended up as waste, accumulating to over 6 billion metric tons (6.6 billion tons) by now.

Of this mountain of plastic waste, only 9% had been recycled, while 12% had been incinerated. The rest was still around, either in landfill sites—some of which may erode in the long term and release their plastic into the open—or in the environment, where a large part of it follows the downhill flow of the hydrological cycle and then accumulates in the oceans. Large quantities of waste are emitted directly into the oceans, from ships and coastlines. On the trends observed, the authors extrapolated that the cumulative amount of plastic waste generated will continue to grow exponentially, quadrupling to more than 25 billion metric tons (27.6 billion tons) by 2050.

What happens to the plastic in the oceans became a major environmental concern in the 2010s. Larger items floating for long periods can transfer species to distant shores, where they may become invasive. Smaller particles, arising either from granulated materials (such as the microbeads still used in many cosmetic products) or from the breakage of items that have become brittle due to the long exposure to sunlight, can enter the food chain. In addition, larger items like plastic bags and discarded fishing gear are known to endanger animals that get entangled.

One particularly vexing problem that has found wide attention is the fact that large seabirds such as albatrosses appear to mistake plastic items for food. Disturbing photos of dead and decaying albatross chicks with their stomachs full of plastic waste, taken by artist Chris Jordan on the Midway Atoll (near Hawaii), alerted the world to this problem when they were first published in 2009.[21]

In 2016, Gabrielle Nevitt's group at the University of California, Davis, presented a surprising, but plausible, explanation of this phenomenon.[22] Albatrosses and other seabirds from the order Procellariiformes ("tube-nosed" birds like petrels) are known to detect the volatile chemical dimethyl sulfide (DMS) and use it as a guide to find prey organisms, such as krill, grazing on phytoplankton. In oceanic ecosystems, these grazers break down dimethylsulfoniopropionate from phytoplankton and release DMS.

The researchers showed experimentally that plastic objects exposed to the marine environment for under a month already suffer biofouling and emit enough DMS to be detectable by seabirds. Thus, seabirds that use DMS as a cue to find food are six times more likely to ingest plastic items than birds that don't. As the amount of waste washed into the oceans continues to increase, this problem is also likely to grow.

Seabirds are among the species most vulnerable to plastic pollution, but the authors warned that other species, including sea turtles, fishes, and some mammals, may also be at risk but haven't been studied sufficiently yet. And even when the effects are studied, it is far from clear what measures could avert the problem—other than reducing the amount of plastic waste present in the environment.

As the "Western" lifestyle, based on cheap fossil fuels and disposable plastics, is being globalized and set to expand to the entire, still-growing world population, it will be hard to dent the exponential growth curve of plastic waste predicted by Geyer and colleagues. Symbolic gestures like bans on disposable shopping bags have had some success regionally. In England, for example, larger shops have been forced to charge five pence for disposable plastic bags since October 2015. As a result, the number of such bags handed out to customers has dropped by 83%.

One measure that is relatively straightforward and can have biological and possibly health benefits is a ban on microbeads in cosmetic products. Political initiatives against some or all the microbeads have already been taken in the United States, Canada, the Netherlands, and Ireland. In 2018, the UK government introduced a ban on microbeads in all products that are meant to wash down the drain immediately, such as shampoos and toothpaste. The ban doesn't cover products designed to stay on the skin for longer, such as makeup or sunscreen.

Other well-established plastic habits will be harder to change, including the packaging of virtually all products in plastic. The resulting waste is not only endangering wildlife but also changing our planet and defining our epoch.

There is a growing consensus that we are now living in the Anthropocene, the new geological time unit characterized by human activities shaping the Earth system. But when did it start? If there are any geologists 20,000 years from now, which markers will they use to define the start of the Anthropocene?

Early markers could be the Industrial Revolution, and the start of the consequent rise of excess carbon dioxide in the atmosphere. However, a mid-twentieth-century onset would capture the "Great Acceleration" of the global economy and the rapid advance of science and technology. These developments have left their marks around the globe, while the impact of the Industrial Revolution was initially limited to certain countries.

Radioactive isotopes distributed globally by nuclear explosions from Hiroshima in 1945 through to the last open-air nuclear tests conducted in the 1960s provide a very clear global geological marker that could be used to define the onset of the Anthropocene. The only drawback is that specialized laboratory equipment is needed to detect these isotopes. Plastic debris, by contrast, has had a more gradual transition but is still an extremely sharp signal on geological timescales and has the advantage of being easily discernible with field methods.

Jan Zalasiewicz from University of Leicester, United Kingdom, and colleagues summarized the case for plastics as a marker of the Anthropocene in a review.[23] The authors concluded that "stratigraphically, plastics within sediments comprise a good practical indicator of Anthropocene strata, using a mid-20th century beginning for this postulated epoch." They acknowledged that the lack of pinpoint precision in the onset of the plastic age means that plastics cannot be expected to act as primary markers for precisely defining the start of the Anthropocene. Future geologists visiting our neighborhood from other stars may find it useful, though, that the plastic signature extends to other Solar System bodies, including the Moon, Mars, and Titan (Saturn's largest moon).

Plastics are also an important part of another new concept used to describe the changes that human activities have made to our planet: the technosphere. As reviewed by Zalasiewicz and colleagues, the physical technosphere is currently estimated to a mass of 30 trillion metric tons (33 trillion tons), or 50 kilograms per square meter (10.2 lb./ft.²), of the entire surface of the Earth (including oceans), or 4,000 metric tons (4,400 tons) per person alive today.[24] It is thus an extension of our species that exceeds our own biomass by five orders of magnitude—and our biomass is already exceptionally large. Humanity's total biomass is estimated to be double the total biomass of large vertebrates existing before our ancestors started hunting them to extinction (see chapter 5).

The currently existing, and rapidly growing, technosphere is thus an outgrowth of life on Earth that is developing comparable stature to the biosphere. Much of it is in solid structures made of concrete of steel, but an increasing part consists of the all-pervasive and mobile plastics. What will become of those when the Anthropocene comes to an end and geology moves on?

As plastics have been around for less than a century, there can be no direct evidence as to what will happen to them on geological timescales. However, geologists like Zalasiewicz have proposed speculative scenarios that are based on the fact that plastics are carbon rich, like the biomass that we observe in fossilized form, and thus may eventually see a similar range of fates as those of organisms of the past.

When our waste becomes buried in sedimentary strata, for instance, Zalasiewicz predicts that some of the hydrocarbons contained in the plastics may be released to become fossil-fuel reserves of the future. Some items may mold the surrounding sediment before being broken down, leaving future researchers from species yet to evolve scratching their heads over the shapes of ballpoint pens and CD cases.

One may find the geologist's view reassuring that, in the long term, everything we have made from the Earth will merge back into it, and our legacy will only testify to a weird phase that our planet went through, like the Ediacaran (when evolution tried many different body plans for animals). In the short term, however, if we want to continue to enjoy the rich but dwindling biodiversity of life on Earth, we really must do something drastic to curb our production of plastics.

Life after the Anthropocene

The year 2020 was our glorious future once. After the millennium, countless predictions and prognoses focused on this appealingly regular number to promise us everything from the decoding of the human brain to asteroid mining and a lunar base, not to forget the inescapable flying cars. Politicians set targets and financiers promised returns, all on the basic assumption that things were going to get better and humanity was on a path to progress.

When 2020 arrived, we woke up to the realization that most of those promises hadn't materialized. Worse still, much of what was bright and new in the early days of the century had turned ugly. The global rollout of information technology and social networks may have brought humanity closer, but it is increasingly causing concerns with its threats of surveillance and the bubble effects causing the polarization of political camps.

After 25 UN-sponsored global climate summits, the global CO_2 emissions were still rising, and companies kept investing in the discovery and extraction of additional amounts of fossil fuels that we really can't afford to burn.

The widespread rise of far-right populism peddling simplistic answers to complex problems, if not ignoring them altogether, isn't helping anybody except those profiteering from multiple crises. All of which was then made even worse by the arrival of the COVID-19 pandemic.

With the single exception of widespread progress in the recognition of LGBTQ rights, the second decade of this century hasn't done much to support the narrative of human progress. Still, futurologists are now pondering what 2050 might bring, while some scientists think further ahead to consider how the biosphere might recover after humans.

As humanity failed to address the global problems of our time in the 2010s, the 2020s are even more crucial for resetting the course. Continued failure to act is likely to make the extinction crisis and catastrophic climate change irreversible, and conditions for humans and wildlife alike could start to deteriorate rapidly.

Thus the different outcomes of the paths we could choose now should be quite dramatically different by 2050. Based on this idea, the think tank Foresight, Research and Innovation, of the international engineering firm Arup, prepared a report with four different scenarios for 2050, which was released in December 2019.[25]

Based on the trends of 2019, the authors analyzed plausible future developments, especially for human societies and the fate of the environment. Their four scenarios are based on the possibility that either or both could improve or decline.

Thus, the scenario modeled on the then current business-as-usual trajectory, labeled "Humans Inc.," predicts that human society progresses but the planetary health declines. Life carries on like it did (as their crystal ball didn't see the pandemic coming), but the continued struggle against the rising sea levels and more frequent extreme weather events is beginning to take its toll.

The "Extinction Express" scenario describes a trajectory in which both the planetary health and human societies deteriorate. In this dystopian future, only the wealthiest can afford to breathe clean air, with their parts of cities shielded under domes. Likewise, clean water and healthy food have become a privilege of the wealthy few, while international co-operation has come to a halt and most states aggressively pursue nationalist interests.

The portmanteau "Greentocracy" defined the scenario where the health of the planet improves but only at the expense of people's individual liberties. To save land for nature restoration, most people must live in high-density cities and rely on synthetic food. Extension of marine protected areas sees cruise ships banned for their environmental impacts, but only in 2042 (too late if you ask me).

The most optimistic scenario, in which both the planetary health and societal conditions improve, envisioned a thriving circular economy, and the loss of biodiversity is brought to a halt. According to this view, the first northern white rhino is due to be rewilded in 2042, and by 2050 human activities will no longer exceed the planetary limits.

This trajectory is referred to as "Post Anthropocene," as if to suggest that the human impact on the planet could be reduced to zero by 2050. Jan Zalasiewicz argued, however, that much of the human impact that is now being considered to define the onset date of the Anthropocene is already irreversible. "The Anthropocene has already produced an indelible stratigraphic record. With huge effort, we might be able to clean the waters of plastics—but we can't clean the plastics, or radionuclides, or changed carbon and nitrogen isotope ratios, or fly ash particles, out of the strata already deposited," Zalasiewicz commented.[26] As we will see below, the Anthropocene is more likely to end after our species disappears.

The authors of the Arup report supported each scenario with a timeline, a portrait of a fictional person living in that world in 2050, as well as an analysis of how well the UN's Sustainable Development Goals are being met in that case.

From where we are standing at the beginning of the 2020s, considering the political situations in the biggest economies, the best-case scenario perhaps looks out of reach, but all of them have plausible elements that could in various combinations come our way soon.

All four scenarios from Arup were optimistic in assuming that our present globalized civilization isn't going to collapse within the next three decades. Considering the current volatility of political leadership, in combination with the continuing existence of overkill nuclear arsenals, it appears that the self-destruction of our civilization, and possibly our entire species, seems to be entirely possible. Even more so at the beginning of 2024 as the war in Ukraine continues unabated and conflicts in the Middle East look set to escalate further.

Many other highly developed civilizations have collapsed in the past. The defining criterion of a society's collapse is the rapid loss of complexity. After the collapse of the Roman Empire, for instance, people may have continued to live in the same places, but the political structures simplified, infrastructure was left to decay, and cultural activity declined.

The British astronomer Martin Rees estimated in his 2003 book *Our Final Hour* (published as *Our Final Century* in the United Kingdom) that our species had a 50-50 chance of surviving the twenty-first century. In a 2018 update, *On the Future: Prospects for Humanity* (2018), Rees advocated to explore the potential of geo-engineering.

The question whether a civilization advanced enough to develop the means of its own destruction can survive in the long term is relevant not only for our own sake but also as a possible answer to the apparent lack of advanced civilizations in the nearby regions of the Universe. As Carl Sagan (1934–96) pointed out, the kind of technology that enables a civilization to conquer space might inevitably bear the seeds of its own destruction.

If our civilization manages to show enough sense to survive the twenty-first century, it may be around for a few more centuries, and humans as a species may persist for a few million years, allowing the Anthropocene to unfold into a proper geological epoch, until a new dramatic change occurs.

No species lives forever, though, and there is no chance that living beings with recognizably human features will be around to watch the Sun turn into a red giant. Estimates for human survival range up to 7.8 million years. As Maggie and David Watson from Charles Sturt University, Australia, argued in a comment piece, the inevitable disappearance of humans from the face of the Earth leaves a substantial amount of geological time to the post-Anthropocene.[27]

Fernando de Sousa Mello and Amâncio César Santos Friaça from the University of São Paulo, Brazil, modeled the co-evolution of geosphere, atmosphere, and biosphere to assess the likely fate of the biosphere on Earth.[28] Their results suggest that the end of life on Earth will come long before the widely known conversion of our Sun into a red giant (in 7.6 billion years) but probably not within a billion years.

De Sousa Mello and Santos Friaça found that, as the Solar System's habitable zone drifts outward, conditions on Earth will become more difficult. Thus, C3 plants (those with a more ancestral form of photosynthesis) are predicted to disappear in 170 million years, and C4 plants (those with a derived form better suitable for warmer climates) no later than 840 million years from now, due to the loss of carbon dioxide from the atmosphere. The demise of plants will also limit the availability of carbon to other kinds of organisms— unless something completely different evolves by that time.

Water loss will not be a dramatic problem on this timescale, but temperatures will be going up. The mean surface temperature of the Earth is predicted to reach 100°C (212°F) in 1.63 billion years, bringing Earth to the edge of habitability, at least for the hardiest kinds of extremophile organisms known today. De Sousa Mello and Santos Friaça described this as the endpoint, the "thermal collapse of the biosphere," although earlier work from Jack O'Malley-James at the University of St Andrews, United Kingdom, suggested the persistence of "swansong biospheres" in sheltered niches populated by unicellular lifeforms—an appealing symmetry to the beginnings of life on Earth.[29]

Based on De Sousa Mello and Santos Friaça's projections, Maggie and David Watson argued that complex life still has at least 800 million years ahead of it after the demise of humans, which exceeds the time elapsed since the emergence of complex body plans in the Ediacaran (around 600 million years ago).[30] Thus, most of the evolution of complex life on Earth is still to come.

As conservation scientists, the Watsons raised the question whether there is anything we can do to help biodiversity in a post-Anthropocene world. Obviously, we cannot preserve individual species for eternity, and the ones we now worry about will disappear with us if not before.

They argued, however, that the survival of the bigger branches of the evolutionary tree, today's phyla, has much better chances and may be influenced by our actions. The logical consequence of that thought runs somewhat against the common thinking of conservation science.

If we want to ensure that a broad branch, like the Felidae, has a future in the post-Anthropocene, there is little chance that conservation of the tiger or the cheetah will help. Conversely, it is the most pervasive species of the branch that we should pin our hopes on. Today's feral cats may become the ancestors of future families of Felidae. Today's rats may breed more rodent species, and today's weeds and insect pests may similarly become the ancestors of future biodiversity. Invasive species, widely seen as a problem for today's biodiversity, may become the seeds of a future biosphere.

This thought of an afterlife on Earth gave the Watsons comfort and positivity in a discipline that often functions as a kind of emergency service. "As a crisis discipline, conservation biology takes a toll on its first responders, routinely confronting us with accelerating extinctions and a society increasingly detached from wildlife and nature."[31]

Championing abundant life forms, like mistletoes (Santalales) and gulls and terns (Laridae), in the hope that these lineages will outlive humans, gave them a feeling that is exceedingly rare in conservation science: hope.

NOTES

CHAPTER 1   Plants and Their Little Helpers

1. Matthew P. Nelsen et al., "No support for the emergence of lichens prior to the evolution of vascular plants," *Geobiol.* 18 (2020): 3–13.
2. Matthew P. Nelsen, email message to author, August 28, 2020.
3. Matthew P. Nelsen et al., "The macroevolutionary dynamics of symbiotic and phenotypic diversification in lichens," *Proc. Natl. Acad. Sci. USA* 117 (2020): 21495–21503.
4. Matthew P. Nelsen, email message to author, August 28, 2020.
5. Jen-Pan Huang et al., "Accelerated diversifications in three diverse families of morphologically complex lichen-forming fungi link to major historical events," *Sci. Rep.* 9 (2019): 8518.
6. François Lutzoni et al., "Contemporaneous radiations of fungi and plants linked to symbiosis," *Nat. Commun.* 9 (2018): 5451.
7. Ming Wang et al., "Blumenols as shoot markers for root symbiosis with arbuscular mycorrhizal fungi," *eLife* 7 (2018): e37093.
8. Roberta Fulthorpe, Adam R. Martin, and Marney E. Isaac, "Root endophytes of coffee (*Coffea arabica*): Variation across climatic gradients and relationships with functional traits," *Phytobiomes J.* 4 (2020): 27–39.
9. C. Pepe-Ranney et al., "Surveying the sweetpotato rhizosphere, endophyte, and surrounding soil microbiomes at two North Carolina farms reveals underpinnings of sweetpotato microbiome community assembly," *Phytobiomes J.* 4 (2020): 75–89.

10. Jose Manuel Martí et al., "Metatranscriptomic dynamics after *Verticillium dahliae* infection and root damage in *Olea europaea*," *BMC Plant Biol.* 20 (2020): 79.

11. Maureen Berg and Britt Koskella, "Nutrient- and dose-dependent microbiome-mediated protection against a plant pathogen," *Curr. Biol.* 28 (2018): 2487–2492.

12. Britt Koskella, email message to author, May 20, 2020.

13. Norma M. Morella, Xuening Zhang, and Britt Koskella, "Tomato seed-associated bacteria confer protection of seedlings against foliar disease caused by *Pseudomonas syringae*," *Phytobiomes J.* 3 (2019): 177–190.

14. Norma M. Morella et al., "Successive passaging of a plant-associated microbiome reveals robust habitat and host genotype-dependent selection," *Proc. Natl. Acad. Sci. USA* 117 (2020): 1148–1159.

15. Tao Chen et al., "A plant genetic network for preventing dysbiosis in the phyllosphere," *Nature* 580 (2020): 653–657.

16. Parris T. Humphrey and Noah K. Whiteman, "Insect herbivory reshapes a native leaf microbiome," *Nat. Ecol. Evol.* 4 (2020): 221–229.

17. Wenke Smets and Britt Koskella, "Microbiome: Insect herbivory drives plant phyllosphere dysbiosis," *Curr. Biol.* 30 (2020): R412–R414.

18. Franziska Eberl et al., "Herbivory meets fungivory: Insect herbivores feed on plant pathogenic fungi for their own benefit," *Ecol. Lett.* 23 (2020): 1073–1084.

19. Nicholas C. Dove et al., "Assembly of the *Populus* microbiome is temporally dynamic and determined by selective and stochastic factors," *mSphere* 6 (2021): e01316-20.

20. Geneviève Lajoie and Steven W. Kembel, "Plant-bacteria associations are phylogenetically structured in the phyllosphere," *Mol. Ecol.* 30 (2021): 5572–5587.

21. Kyle M. Meyer et al., "Plant neighborhood shapes diversity and reduces interspecific variation of the phyllosphere microbiome," *ISME Journal* 16 (2022): 1376–1387.

22. Fantin Mesny et al., "Genetic determinants of endophytism in the *Arabidopsis* root mycobiome," *Nat. Commun.* 12 (2021): 7227.

23. Kenneth Dumack et al., "A call for research: A resource of core microbial symbionts of the *Arabidopsis thaliana* microbiome ready and awaiting experimental exploration," *Phytobiomes J.* 5 (2021): 362–366.

24. Christine M. Vogel et al., "Protective role of the *Arabidopsis* leaf microbiota against a bacterial pathogen," *Nat. Microbiol.* 6 (2021): 1537–1548.

25. Nick C. Snelders et al., "An ancient antimicrobial protein co-opted by a fungal plant pathogen for in planta mycobiome manipulation," *Proc. Natl. Acad. Sci. USA* 118 (2021): e2110968118.

26. Ahmed Abdelfattah et al., "Evidence for host–microbiome co-evolution in apple," *New Phytol.* 234 (2022): 2088–2100.

27. David F. Rhoades, "Responses of Alder and Willow to Attack by Tent Caterpillars and Webworms: Evidence for Pheromonal Sensitivity of Willows," in *Plant Resistance to Insects*, ed. Paul A. Hedin (Washington, DC: American Chemical Society, 1983), 55–68.

28. Richard Karban et al., "Deciphering the language of plant communication: Volatile chemotypes of sagebrush," *New Phytol.* 204 (2014): 380–385.

29. Björn Bohman et al., "Discovery of pyrazines as pollinator sex pheromones and orchid semiochemicals: Implications for the evolution of sexual deception," *New Phytol.* 203 (2014): 939–952.

30. Birgit Oelschlägel et al., "The betrayed thief—the extraordinary strategy of *Aristolochia* rotunda to deceive its pollinators," *New Phytol.* 206 (2015): 342–353.

31. Vittorio Venturi and Christoph Keel, "Signaling in the rhizosphere," *Trends Plant Sci.* 21 (2016): 187–198.

32. Monica Gagliano et al., "Experience teaches plants to learn faster and forget slower in environments where it matters," *Oecologia* 175 (2014): 63–72.

33. Jennifer Böhm et al., "The Venus flytrap *Dionaea muscipula* counts prey-induced action potentials to induce sodium uptake," *Curr. Biol.* 26 (2016): 286–295.

CHAPTER 2 Fantastic Animals

1. Roberto Feuda et al., "Improved modeling of compositional heterogeneity supports sponges as sister to all other animals," *Curr. Biol.* 27 (2017): 3864–3870.

2. J. Alex Zumberge et al., "Demosponge steroid biomarker 26-methylstigmastane provides evidence for Neoproterozoic animals," *Nat. Ecol. Evol.* 2 (2018): 1709–1714.

3. L. K. Law et al., "Description and distribution of *Desmacella hyalina* sp. nov. (Porifera, Desmacellidae), a new cryptic demosponge in glass sponge reefs from the western coast of Canada," *Mar. Biodivers.* 50 (2020): 55.

4. Vanessa Schoeppler et al., "Shaping highly regular glass architectures: A lesson from nature," *Sci. Adv.* 3 (2017): eaao2047.

5. Elsa B. Girard et al., "Sponges as bioindicators for microparticulate pollutants?," *Environ. Pollut.* 268 (2021): 115851.

6. Stefano Mariani et al., "Sponges as natural environmental DNA samplers," *Curr. Biol.* 29 (2019): R401–R402.

7. Georg Steinert et al., "Compositional and quantitative insights into bacterial and archaeal communities of South Pacific deep-sea sponges (Demospongiae and Hexactinellida)," *Front. Microbiol.* 11 (2020): 00716.

8. Sergio Vargas, Laura Leiva, and Gert Wörheide, "Short-term exposure to high-temperature water causes a shift in the microbiome of the common aquarium sponge *Lendenfeldia chondrodes*," *Microb. Ecol.* 81 (2021): 213–222.

9. Aimee E. Nicholson-Jack et al., "A hitchhiker guide to manta rays: Patterns of association between *Mobula alfredi*, *M. birostris*, their symbionts, and other fishes in the Maldives," *PLoS ONE* 16 (2021): e0253704.

10. Hugo Lassauce et al., "Diving behavior of the reef manta ray (*Mobula alfredi*) in New Caledonia: More frequent and deeper night-time diving to 672 meters," *PLoS ONE* 15 (2020): e0228815.

11. E. E. Becerril-García et al., "Presence of *Remora remora* on *Mobula birostris* in Revillagigedo National Park, Mexico," *Mar. Freshw. Res.* 71 (2019): 414–417.

12. Karly E. Cohen et al., "Knowing when to stick: Touch receptors found in the remora adhesive disc," *R. Soc. Open Sci.* 7 (2020): 190990.

13. Brooke Flammang, email message to author, August 4, 2021.

14. Brooke E. Flammang et al., "Remoras pick where they stick on blue whales," *J. Exp. Biol.* 223 (2020): jeb226654.

15. Kaelyn M. Gamel, Austin M. Garner, and Brooke E. Flammang, "Bioinspired remora adhesive disc offers insight into evolution," *Bioinspir. Biomim.* 14 (2019): 056014.

16. Ryota Kawanishi and Shinpei Ohashi, "First record of the rare parasitic isopod *Elthusa splendida* (Cymothoidae) from the Pacific

Ocean, based on a specimen found in a museum shark collection," *Species Divers.* 25 (2020): 343–348.

17. Alice R. Jones et al., "Identifying island safe havens to prevent the extinction of the world's largest lizard from global warming," *Ecol. and Evol.* 10 (2020): 10492–10507.

18. Carlos J. Pavón-Vázquez, Ian G. Brennan, and J. Scott Keogh, "A comprehensive approach to detect hybridization sheds light on the evolution of Earth's largest lizards," *Syst. Biol.* 70 (2021): 877–890.

19. Sergio Guerrero-Sanchez et al., "The critical role of natural forest as refugium for generalist species in oil palm-dominated landscapes," *PLoS ONE* 16 (2021): e0257814.

20. Sarah L. Whiteley et al., "Two transcriptionally distinct pathways drive female development in a reptile with both genetic and temperature dependent sex determination," *PLoS Genet.* 17 (2021): e1009465.

21. Neil J. Gemmell et al., "The tuatara genome reveals ancient features of amniote evolution," *Nature* 584 (2020): 403–409.

22. Frédéric Delsuc et al., "Resolving the phylogenetic position of Darwin's extinct ground sloth (*Mylodon darwinii*) using mitogenomic and nuclear exon data," *Proc. R. Soc. B* 285 (2018): 20180214.

23. Frédéric Delsuc et al., "Ancient mitogenomes reveal the evolutionary history and biogeography of sloths," *Curr. Biol.* 29 (2019): 2031–2042.

24. Samantha Presslee et al., "Palaeoproteomics resolves sloth relation-ships," *Nat. Ecol. Evol.* 3 (2019): 1121–1130.

25. Jonathan N. Pauli et al., "Arboreal folivores limit their energetic output, all the way to slothfulness," *Am. Nat.* 188 (2016): 196–204.

26. Mario F. Garcés-Restrepo, M. Zachariah Peery, and Jonathan N. Pauli, "The demography of a resource specialist in the tropics: Cecropia trees and the fitness of three-toed sloths," *Proc. R. Soc. B* 286 (2019): 20182206.

27. Mario F. Garcés-Restrepo, Jonathan N. Pauli, and M. Zachariah Peery, "Natal dispersal of tree sloths in a human-dominated landscape: Implications for tropical biodiversity conservation," *J. Appl. Ecol.* 55 (2018): 2253–2262.

28. Jonathan N. Pauli et al., "A syndrome of mutualism reinforces the lifestyle of a sloth," *Proc. Biol. Sci.* 281 (2014): 20133006.

29. Jens C. Koblitz et al., "Highly directional sonar beam of narwhals (*Monodon monoceros*) measured with a vertical 16 hydrophone array," *PLoS ONE* 11 (2016): e0162069.

30. Greg W. Rouse, Josefin Stiller, and Nerida G. Wilson, "First live records of the ruby seadragon (*Phyllopteryx dewysea*, Syngnathidae)," *Mar. Biodivers. Rec.* 10 (2017): 2.

31. Qiang Lin et al., "The seahorse genome and the evolution of its specialized morphology," *Nature* 540 (2016): 395–399.

32. Otto M. P. Oliveira et al., "Census of Cnidaria (Medusozoa) and Ctenophora from South American marine waters," *Zootaxa* 4194 (2016): 1–256.

33. Liz Langley, "7 amazing animal organs people don't have," *National Geographic*, January 14, 2017, https://www.nationalgeographic.com /science/article/organs-mesentery-snakes-sharks.

CHAPTER 3 Insects Rule the World

1. Joachim Offenberg and Christian Damgaard, "Ants suppressing plant pathogens: A review," *Oikos* 128 (2019): 1691–1703.

2. Matthew P. Nelsen, Richard H. Ree, and Corrie S. Moreau, "Ant-plant interactions evolved through increasing interdependence," *Proc. Natl. Acad. Sci. USA* 115 (2018): 12253–12258.

3. Katrina M. Kaur et al., "Using text-mined trait data to test for cooperate-and-radiate co-evolution between ants and plants," *PLoS Comp. Biol.* 15 (2019): e1007323.

4. PLoS, "Ant-plant partnerships may play unexpected role in ant evolution," EurekAlert! news release, October 3, 2019, https://www .eurekalert.org/news-releases/622322.

5. Sabrina Simon et al., "Old World and New World Phasmatodea: Phylogenomics resolve the evolutionary history of stick and leaf insects," *Front. Ecol. Evol.* 7 (2019): 345.

6. Dominic A. Evangelista et al., "An integrative phylogenomic approach illuminates the evolutionary history of cockroaches and termites (Blattodea)," *Proc. R. Soc. B* 286 (2019): 20182076.

7. Ales Bucek et al., "Evolution of termite symbiosis informed by transcriptome-based phylogenies," *Curr. Biol.* 29 (2019): 3728–3734.

8. Benjamin Wipfler et al., "Evolutionary history of Polyneoptera and its implications for our understanding of early winged insects," *Proc. Natl. Acad. Sci. USA* (2019): 3024–3029.

9. Grace K. Charles et al., "Termite mound cover and abundance respond to herbivore-mediated biotic changes in a Kenyan savanna," *Ecol. Evol.* 11 (2021): 7226–7238.

10. Juan A. Bonachela et al., "Termite mounds can increase the robustness of dryland ecosystems to climatic change," *Science* 347 (2015): 651–655.

11. Corina E. Tarnita et al., "A theoretical foundation for multi-scale regular vegetation patterns," *Nature* 541 (2017): 398–401.

12. Morgan Kelly, "In African 'fairy circles,' a template for nature's many patterns," Princeton University, January 19, 2017, https://www .princeton.edu/news/2017/01/19/african-fairy-circles-template-natures -many-patterns.

13. Andrés Arenas and Flavio Roces, "Avoidance of plants unsuitable for the symbiotic fungus in leaf-cutting ants: Learning can take place entirely at the colony dump," *PLOS ONE* 12 (2017): e0171388.

14. Tongchuan Li, Ming'an Shao, and Yuhua Jia, "Characteristics of soil evaporation and temperature under aggregate mulches created by burrowing ants (*Camponotus japonicus*)," *Soil Sci. Soc. Am. J.* 81 (2017): 259–267.

15. Martin Nyffeler and Klaus Birkhofer, "An estimated 400–800 million tons of prey are annually killed by the global spider community," Sci. Nat. 104 (2017): 30.

16. Bernhard Misof et al., "Phylogenomics resolves the timing and pattern of insect evolution," *Science* 346 (2014): 763–767.

17. Chikako Ishida, Masumi Kono, and Shoko Sakai, "A new pollination system: Brood-site pollination by flower bugs in *Macaranga* (Euphorbiaceae)," *Ann. Bot.* 103 (2009): 39–44.

18. Joana Couto et al., "Antiplasmodial activity of tick defensins in a mouse model of malaria," *Ticks Tick-Borne Dis.* 9 (2018): 844–849.

19. Leopold Salzenstein, "East Africa deploys huge volumes of 'highly hazardous' pesticides against locust plague," Mongabay, April 1, 2021, https://news.mongabay.com/2021/04/east-africa-deploys-huge-volumes -of-highly-hazardous-pesticides-against-locust-plague/.

20. Y. Golov et al., "Sexual behavior of the desert locust during intra- and inter-phase interactions," *J. Insect Behav.* 31 (2018): 629–641.

21. Amir Ayali, email message to author, May 5, 2021.

22. Omer Lavy et al., "Locust bacterial symbionts: An update," *Insects* 11 (2020): 655.

23. Omer Lavy et al., "Microbiome-related aspects of locust density-dependent phase transition," *Environ. Microbiol.* 24 (2022): 507–516.

24. Omer Lavy et al., "Dynamics of bacterial composition in the locust reproductive tract are affected by the density-dependent phase," *FEMS Microbiol. Ecol.* 96 (2020): fiaa044.

25. Daniel Knebel et al., "Collective motion as a distinct behavioral state of the individual," *iScience* 24 (2021): 102299.

26. Arianne J. Cease et al., "Heavy livestock grazing promotes locust outbreaks by lowering plant nitrogen content," *Science* (2012): 467–469.

27. Marion Le Gall, Rick Overson, and Arianne Cease, "A global review on locusts (Orthoptera: Acrididae) and their interactions with livestock grazing practices," *Front. Ecol. Evol.* 7 (2019): 00263.

28. Marion Le Gall et al., "Nitrogen fertilizer decreases survival and reproduction of female locusts by increasing plant protein to carbohydrate ratio," *J. Anim. Ecol.* 89 (2020): 2214–2221.

29. Stav Talal et al., "Plant carbohydrate content limits performance and lipid accumulation of an outbreaking herbivore," *Proc. R. Soc. B.* 287 (2020): 20202500.

30. Amir Ayali, email message to author, May 5, 2021.

CHAPTER 4 Looking After Our Forests

1. J. W. Ferry Slik et al., "Phylogenetic classification of the world's tropical forests," *Proc. Natl. Acad. Sci. USA* 115 (2018): 1837–1842.

2. Michael D. Pirie et al., "Parallel diversifications of *Cremastosperma* and *Mosannona* (Annonaceae), tropical rainforest trees tracking Neogene upheaval of South America," *R. Soc. Open Sci.* 5 (2018): 171561.

3. Sandra Díaz et al., "The global spectrum of plant form and function," *Nature* 529 (2016): 167–171.

4. Zeqing Ma et al., "Evolutionary history resolves global organization of root functional traits," *Nature* 555 (2018): 94–97.

5. Franziska Taubert et al., "Global patterns of tropical forest fragmentation," *Nature* 554 (2018): 519–522.

6. James E. M. Watson et al., "The exceptional value of intact forest ecosystems," *Nat. Ecol. Evol.* 2 (2018): 599–610.

7. José M. Capriles et al., "Persistent early to middle Holocene tropical foraging in southwestern Amazonia," *Sci. Adv.* 5 (2019): eaav5449.

8. Penn State, "Human settlements in Amazonia much older than previously thought," EurekAlert! news release, April 24, 2019, https://www.eurekalert.org/news-releases/625393.

9.   Neil A. Duncan et al., "Pre-Columbian fire management and control of climate-driven floodwaters over 3,500 years in southwestern Amazonia," *Proc. Natl. Acad. Sci. USA* 118 (2021): e2022206118.

10. Umberto Lombardo et al., "Early Holocene crop cultivation and landscape modification in Amazonia," *Nature* 581 (2020): 190–193.

11. Alexander Koch et al., "Earth system impacts of the European arrival and Great Dying in the Americas after 1492," *Quat. Sci. Rev.* 207 (2019): 13–36.

12. Dolores R. Piperno et al., "A 5,000-year vegetation and fire history for *tierra firme* forests in the Medio Putumayo-Algodón watersheds, northeastern Peru," *Proc. Natl. Acad. Sci. USA* 118 (2021): e2022213118.

13. Frederick Draper et al., "Amazon tree dominance across forest strata," *Nat. Ecol. Evol.* 5 (2021): 757–767.

14. Piperno et al., "A 5,000-year vegetation and fire history."

15. Jonathan Jones, "'Paradise exists!': Sebastião Salgado's stunning voyage into Amazônia," *Guardian*, June 21, 2021, https://www .theguardian.com/artanddesign/2021/jun/21/paradise-exists -sebastiao-salgados-stunning-voyage-into-amazonia.

16. Martin J. P. Sullivan et al., "Long-term thermal sensitivity of Earth's tropical forests," *Science* 368 (2020): 869–874.

17. Martin Sullivan, email message to author, June 3, 2020.

18. Matthew C. Hansen et al., "The fate of tropical forest fragments," *Sci. Adv.* 6 (2020): eaax8574.

19. Nate G. McDowell et al., "Pervasive shifts in forest dynamics in a changing world," *Science* 368 (2020): eaaz9463.

20. University of Birmingham, "Global environmental changes are leading to shorter, younger trees—new study," EurekAlert! news release, May 28, 2020, https://www.eurekalert.org/news-releases/460344.

21. Lander Baeten et al., "Identifying the tree species compositions that maximize ecosystem functioning in European forests," *J. Appl. Ecol.* 56 (2019): 733–744.

22. William F. Laurance et al., "An Amazonian rainforest and its fragments as a laboratory of global change," *Biol. Rev.* 93 (2017): 223–247.

23. Ricardo Rocha et al., "Consequences of a large-scale fragmentation experiment for Neotropical bats: Disentangling the relative importance of local and landscape-scale effects," *Landscape Ecol.* 32 (2017): 31–45.

24. Johannes Pirker et al., "What are the limits to oil palm expansion?," *Global Environ. Change* 40 (2016): 73–81.

25. Jingjing Liang et al., "Positive biodiversity-productivity relationship predominant in global forests," *Science* 354 (2016): aaf8957.

26. Donal P. McCarthy et al., "Financial costs of meeting global biodiversity conservation targets: Current spending and unmet needs," *Science* 338 (2012): 946–949.

27. DRYFLOR et al., "Plant diversity patterns in neotropical dry forests and their conservation implications," *Science* 353 (2016): 1383–1387.

28. D. Nogués-Bravo et al., "Amplified plant turnover in response to climate change forecast by Late Quaternary records," *Nature Clim. Change* 6 (2016): 1115–1119.

29. Helen Ding et al., "Climate benefits, tenure costs: The economic case for securing indigenous land rights in the Amazon," World Resources Institute, October 16, 2016, http://www.wri.org/publication/climate -benefits-tenure-costs.

CHAPTER 5   This Time, the Asteroid Is Us

1. A. Benítez-López et al., "The impact of hunting on tropical mammal and bird populations," *Science* 356 (2017): 180–183.

2. Ana Benítez-López et al., "Intact but empty forests? Patterns of hunting-induced mammal defaunation in the tropics," *PLoS Biol.* 17 (2019): e3000247.

3. Radboud University Nijmegen, "Scientists estimate: Half of tropical forests under hunting pressure," EurekAlert! news release, May 14, 2019, https://www.eurekalert.org/news-releases/631469.

4. Ana Benítez-López et al., "Intact but empty forests?"

5. Robert S. C. Cooke, Felix Eigenbrod, and Amanda E. Bates, "Projected losses of global mammal and bird ecological strategies," *Nat. Commun.* 10 (2019): 2279.

6. Frédéric Jiguet et al., "Unravelling migration connectivity reveals unsustainable hunting of the declining ortolan bunting," *Sci. Adv.* 5 (2019): eaau2642.

7. Jason J. Gregg, "In Malta, legal loopholes give poachers cover to hunt migratory birds," Mongabay, January 11, 2019, https://news.mongabay .com/2019/01/in-malta-legal-loopholes-give-poachers-cover-to-hunt -migratory-birds/.

8. J. Berton C. Harris et al., "Measuring the impact of the pet trade on Indonesian birds," *Conserv. Biol.* 31 (2017): 394–405.

9. Thomas S. Kraft et al., "The energetics of uniquely human subsistence strategies," *Science* 374 (2021): eabf0130.

10. Nicholas B. Holowka et al., "Forest terrains influence walking kinematics among Indigenous Tsimane of the Bolivian Amazon," *Evol. Hum. Sci.* 4 (2022): e19.

11. University of California, Santa Barbara, "Anthropologists study the energetics of uniquely human subsistence strategies," ScienceDaily news release, January 3, 2022, https://sciencedaily.com/releases/2022/01/220103145553.htm.

12. Marcus J. Hamilton, "Collective computation, information flow, and the emergence of hunter-gatherer small-worlds," *J. Social Comput.* 3 (2022): 18–37.

13. Nina Marchi et al., "The genomic origins of the world's first farmers," *Cell* 185 (2022): 1842–1859.

14. Morten E. Allentoft et al., "Population genomics of post-glacial western Eurasia," *Nature* 625 (2024): 301–311.

15. Kraft et al., "The energetics of uniquely human subsistence strategies."

16. Alexander Nater et al., "Morphometric, behavioral, and genomic evidence for a new orangutan species," *Curr. Biol.* 27 (2017): 3487–3498.

17. Hans Nicholas Jong, "Protests flare as pressure mounts on dam project in orangutan habitat," Mongabay, March 1, 2019, https://news.mongabay.com/2019/03/protests-flare-as-pressure-mounts-on-dam-project-in-orangutan-habitat/.

18. Bill Laurance, email message to author, March 11, 2019.

19. Serge A. Wich et al., "Land-cover changes predict steep declines for the Sumatran orangutan (*Pongo abelii*)," *Sci. Adv.* 2 (2016): e1500789.

20. Malachi Chadwick, "Rang-tan: Watch the film," Greenpeace, May 11, 2019, https://www.greenpeace.org.uk/news/watch-rang-tan-film/.

21. Maria Voigt et al., "Global demand for natural resources eliminated more than 100,000 Bornean orangutans," *Curr. Biol.* 28 (2018): 761–769.

22. Isabelle B. Laumer et al., "Orangutans (*Pongo abelii*) make flexible decisions relative to reward quality and tool functionality in a multi-dimensional tool-use task," *PLoS ONE* 14 (2019): e0211031.

23. Stephanie N. Spehar et al., "Orangutans venture out of the rainforest and into the Anthropocene," *Sci. Adv.* 4 (2018): e1701422.

24. Ridwan Setiawan et al., "Preventing global extinction of the Javan rhino: Tsunami risk and future conservation direction," *Conserv. Lett.* 11 (2017): e12366.

25. Cindy Harper et al., "Robust forensic matching of confiscated horns to individual poached African rhinoceros," *Curr. Biol.* 28 (2018): R13–R14.

26. Frederick Chen, "The economics of synthetic rhino horns," *Ecol. Econ.* 141 (2017): 180–189.

27. Wake Forest University, "Synthetic horns may save rhinos if they are not like the real thing," EurekAlert! news release, July 5, 2017, https://www.eurekalert.org/news-releases/832216.

28. Yoshan Moodley et al., "Extinctions, genetic erosion and conservation options for the black rhinoceros (*Diceros bicornis*)," *Sci. Rep.* 7 (2017): 41417.

29. Herman L. Mays Jr. et al., "Genomic analysis of demographic history and ecological niche modeling in the endangered Sumatran rhinoceros *Dicerorhinus sumatrensis*," *Curr. Biol.* 28 (2018): 70–76.

30. Robert H. Cowie, Philippe Bouchet, and Benoît Fontaine, "The sixth mass extinction: Fact, fiction or speculation?," *Biol. Rev.* 97 (2022): 640–663.

31. University of Hawai'i at Mānoa, "Strong evidence shows Sixth Mass Extinction of global biodiversity in progress," EurekAlert! news release, January 13, 2022, https://www.eurekalert.org/news-releases /940163.

32. James R. Allan et al., "The minimum land area requiring conservation attention to safeguard biodiversity," *Science* 376 (2022): 1094–1101.

33. Chris T. Darimont et al., "The unique ecology of human predators," *Science* 349 (2015): 858–860.

34. Food and Agriculture Organization of the United Nations (FAO), "Food balances (2017)," FAOSTAT, accessed February 6, 2024, https://www.fao.org/faostat/en/#data/FBS.

35. Ophelia Deroy, "Eat insects for fun, not to help the environment," *Nature* 521 (2015): 395.

36. CSIRO, "An industry with legs: Australia's first edible insects," news release, April 29, 2021, https://www.csiro.au/en/news/All/News/2021 /April/An-industry-with-legs-Australias-first-edible-insects-roadmap.

CHAPTER 6 Save Our Seas

1. Willem Renema et al. "Are coral reefs victims of their own past success?," *Sci. Adv.* 2 (2016): e1500850.

2. Terry P. Hughes et al., "Global warming and recurrent mass bleaching of corals," *Nature* 543 (2017): 373–377.

3. Timothy D. Swain et al., "Coral bleaching response index: A new tool to standardize and compare susceptibility to thermal bleaching," *Glob. Change Biol.* 22 (2016): 2475–2488.

4. Tracy D. Ainsworth et al., "Climate change disables coral bleaching protection on the Great Barrier Reef," *Science* 352 (2016): 338–342.

5. Elaine Baker, Kimberly A. Puglise, and Peter T. Harris, eds., "Mesophotic coral ecosystems: A lifeboat for coral reefs?," UN Environment Programme, July 5, 2016, https://www.unep.org/resources/report/mesophotic-coral-ecosystems-lifeboat-coral-reefs.

6. Elena Couce, Andy Ridgwell, and Erica J. Hendy, "Future habitat suitability for coral reef ecosystems under global warming and ocean acidification," *Glob. Change Biol.*, 19 (2013): 3592–3606.

7. Madeleine J. H. van Oppen et al., "Building coral reef resilience through assisted evolution," *Proc. Natl. Acad. Sci. USA* 112 (2015): 2307–2313.

8. Rio Kashimoto et al., "Transcriptomes of giant sea anemones from Okinawa as a tool for understanding their phylogeny and symbiotic relationships with anemonefish," *Zool. Sci.* 39 (2022): 374–387.

9. Robert H. MacArthur and Edward O. Wilson, *The Theory of Island Biogeography* (Princeton, NJ: Princeton University Press, 1967).

10. Shana K. Goffredi et al., "Hydrothermal vent fields discovered in the southern Gulf of California clarify role of habitat in augmenting regional diversity," *Proc. R. Soc. B* 284 (2017): 20170817.

11. Hudson T. Pinheiro et al., "Island biogeography of marine organisms," *Nature* 549 (2017): 82–85.

12. Sean M. Evans et al., "Patterns of species range evolution in Indo-Pacific reef assemblages reveal the Coral Triangle as a net source of transoceanic diversity," *Biol. Lett.* 12 (2016): 20160090.

13. Cristian Salinas et al., "Seagrass losses since mid-20th century fuelled CO_2 emissions from soil carbon stocks," *Glob. Change Biol.* 26 (2020): 4772–4784.

14. Edith Cowan University, "New study sparks fresh call for seagrass preservation," EurekAlert! news release, July 7, 2020, https://www.eurekalert.org/news-releases/555294.

15. Rebecca K. James et al., "Maintaining tropical beaches with seagrass and algae: A promising alternative to engineering solutions," *BioScience* 69 (2019): 136–142.

16. Royal Netherlands Institute for Sea Research, "Seagrass saves beaches and money," EurekAlert! news release, January 2, 2019, https://www.eurekalert.org/news-releases/500726.

17. R. K. James et al., "Tropical biogeomorphic seagrass landscapes for coastal protection: Persistence and wave attenuation during major storms events," *Ecosystems* 24 (2021): 301–318.

18. A. Challen Hyman et al., "Long-term persistence of structured habitats: Seagrass meadows as enduring hotspots of biodiversity and faunal stability," *Proc. Roy. Soc. B* 286 (2019): 20191861.

19. Richard K. F. Unsworth, Lina Mtwana Nordlund, and Leanne C. Cullen-Unsworth, "Seagrass meadows support global fisheries production," *Conserv. Lett.* 12 (2019): e12566.

20. Jeffrey J. Kelleway et al., "A national approach to greenhouse gas abatement through blue carbon management," *Glob. Env. Change* 63 (2020): 102083.

21. Peter I. Macreadie et al., "The future of blue carbon science," *Nat. Commun.* 10 (2019): 3998.

22. Phillip A. Morin et al., "Building genomic infrastructure: Sequencing platinum-standard reference-quality genomes of all cetacean species," *Mar. Mam. Sci.* 36 (2020): 1356–1366.

23. Phillip A. Morin et al., "Reference genome and demographic history of the most endangered marine mammal, the vaquita," *Mol. Ecol. Resour.* 21 (2021): 1008–1020.

24. Andrew D. Foote et al., "Killer whale genomes reveal a complex history of recurrent admixture and vicariance," *Mol. Ecol.* 28 (2019): 3427–3444.

25. Michael V. Westbury et al., "Narwhal genome reveals long-term low genetic diversity despite current large abundance size," *iScience* 15 (2019): 592–599.

26. Patricia E. Rosel et al., "A new species of baleen whale (*Balaenoptera*) from the Gulf of Mexico, with a review of its geographic distribution," *Mar. Mam. Sci.* 37 (2021): 577–610.

27. Alexandre N. Zerbini et al., "Assessing the recovery of an Antarctic predator from historical exploitation," *R. Soc. Open Sci.* 6 (2019): 190368.

CHAPTER 7 Living with Animals

1. Thomas Cucchi et al., "Tracking the Near Eastern origins and European dispersal of the western house mouse," *Sci. Rep.* 10 (2020): 8276.
2. Robin G. Allaby et al., "Geographic mosaics and changing rates of cereal domestication," *Phil. Trans. R. Soc. B* 372 (2017): 20160429.
3. Cucchi et al., "Tracking the Near Eastern origins."
4. Claudio Ottoni et al., "The palaeogenetics of cat dispersal in the ancient world," *Nat. Ecol. Evol.* 1 (2017): 0139.
5. Evelyn T. Todd et al., "The genomic history and global expansion of domestic donkeys," *Science* 377 (2022): 1172–1180.
6. Pablo Librado et al., "The origins and spread of domestic horses from the western Eurasian steppes," *Nature* 598 (2021): 634–640.
7. E. Andrew Bennett et al., "The genetic identity of the earliest human-made hybrid animals, the kungas of Syro-Mesopotamia," *Sci. Adv.* 8 (2022): eabm0218.
8. Karen C. Seto, Burak Güneralp, and Lucy R. Hutyra, "Global forecasts of urban expansion to 2030 and direct impacts on biodiversity and carbon pools," *Proc. Natl. Acad. Sci. USA* 109 (2012): 16083–16088.
9. Rob Dunn, email message to author, May 16, 2018.
10. Misha Leong et al., "The habitats humans provide: Factors affecting the diversity and composition of arthropods in houses," *Sci. Rep.* 7 (2017): 15347.
11. Arjen E. van't Hof et al., "The industrial melanism mutation in British peppered moths is a transposable element," *Nature* 534 (2016): 102–105.
12. Elizabeth M. A. Kern and R. Brian Langerhans, "Urbanization drives contemporary evolution in stream fish," *Glob. Change Biol.* 24 (2018): 3791–3803.
13. Panagiotis Theodorou et al., "Genome-wide single nucleotide polymorphism scan suggests adaptation to urbanization in an important pollinator, the red-tailed bumblebee (*Bombus lapidarius* L.)," *Proc. R. Soc. B.* 285 (2018): 20172806.
14. Martin-Luther-Universität Halle-Wittenberg, "Evolution: Urban life leaves behind traces in the genome of bumblebees," EurekAlert! news release, April 18, 2018, https://www.eurekalert.org/news-releases/480393.
15. Marina Alberti et al., "Global urban signatures of phenotypic change in animal and plant populations," *Proc. Natl. Acad. Sci. USA* 114 (2017): 8951–8956.

16. Marc T. J. Johnson and Jason Munshi-South, "Evolution of life in urban environments," *Science* 358 (2017): eaam8327.

17. Janis M. Wolf et al., "Urban affinity and its associated traits: A global analysis of bats," *Glob. Change Biol.* 28 (2022): 5667–5682.

18. Barbara C. Klump et al., "Innovation and geographic spread of a complex foraging culture in an urban parrot," *Science* 373 (2021): 456–460.

19. James S. Santangelo et al., "Global urban environmental change drives adaptation in white clover," *Science* 375 (2022): 1275–1281.

20. L. Ruth Rivkin and Marc T. J. Johnson, "The impact of urbanization on outcrossing rate and population genetic variation in the native wildflower, *Impatiens capensis*," *J. Urb. Ecol.* 8 (2022): juac009.

21. O. V. Sanderfoot, J. D. Kaufman, and B. Gardner, "Drivers of avian habitat use and detection of backyard birds in the Pacific Northwest during COVID-19 pandemic lockdowns," *Sci. Rep.* 12 (2022): 12655.

22. L. Ruth Rivkin et al., "A roadmap for urban evolutionary ecology," *Evol. Appl.* 12 (2019): 384–398.

23. Ramya Nair et al., Dhee, "Sharing spaces and entanglements with big cats: The Warli and their Waghoba in Maharashtra, India," *Front. Conserv. Sci.* 2 (2021): 683356.

24. Wildlife Conservation Society, "Study: How a large cat deity helps people to share space with leopards in India," EurekAlert! news release, July 8, 2021, https://www.eurekalert.org/news-releases/912849.

25. Phuntsho Thinley et al., "Understanding human–canid conflict and coexistence: Socioeconomic correlates underlying local attitude and support toward the endangered dhole (*Cuon alpinus*) in Bhutan," *Front. Conserv. Sci.* 2 (2021): 691507.

26. Félix Landry Yuan et al., "Sacred groves and serpent-gods moderate human–snake relations," *People Nat.* 2 (2020): 111–122.

27. Lynne R. Baker, Adebowale A. Tanimola, and Oluseun S. Olubode, "Complexities of local cultural protection in conservation: The case of an endangered African primate and forest groves protected by social taboos," *Oryx* 52 (2018): 262–270.

28. Holly K. Nesbitt et al., "Collective factors reinforce individual contributions to human-wildlife coexistence," *J. Wildl. Manag.* 85 (2021): 1280–1295.

29. Clayton T. Lamb et al., "The ecology of human–carnivore coexistence," *Proc. Natl. Acad. Sci. USA* 117 (2020): 17876–17883.

CHAPTER 8   Listen to Nature

1. Gabriel Jorgewich-Cohen et al., "Common evolutionary origin of acoustic communication in choanate vertebrates," *Nat. Commun.* 13 (2022): 6089.

2. Zhuo Chen and John J. Wiens, "The origins of acoustic communication in vertebrates," *Nat. Commun.* 11 (2020): 369.

3. Jorgewich-Cohen et al., "Common evolutionary origin of acoustic communication in choanate vertebrates."

4. Claudia Lacroix, Christina Davy, and Njal Rollinson, "Hatchling vocalizations and beneficial social interactions in subterranean nests of a widespread reptile," *Anim. Behav.* 187 (2022): 233–244.

5. Isabelle Charrier et al., "First evidence of underwater vocalizations in green sea turtles *Chelonia mydas*," *Endanger. Species Res.* 48 (2022): 31–41.

6. Ajay Bedi et al., "Wild observations of the reproductive behaviour and first evidence of vocalization in crocodile newt *Tylototriton himalayanus* (Caudata: Salamandridae) from the Himalayan biodiversity hotspot in Eastern India," *Salamandra* 57 (2021): 65–74.

7. A. Staniewicz et al., "Courtship and underwater communication in the Sunda gharial (*Tomistoma schlegelii*)," *Bioacoustics* 31 (2022): 435–449.

8. Marcus Thadeu T. Santos et al., "Complex acoustic signals in *Crossodactylodes* (Leptodactylidae, Paratelmatobiinae): A frog genus historically regarded as voiceless," *Bioacoustics* 31 (2022): 175–190.

9. Andréa Thiebault et al., "First evidence of underwater vocalisations in hunting penguins," *Peer J.* 7 (2019): e8240.

10. Emily Doolittle, "Birdsong and music," *Curr. Biol.* 32 (2022): R1064–R1066.

11. Melissa J. Coleman et al., "Neurophysiological coordination of duet singing," *Proc. Natl. Acad. Sci. USA* 118 (2021): e2018188118.

12. Susanne Hoffmann et al., "Duets recorded in the wild reveal that interindividually coordinated motor control enables cooperative behavior," *Nat. Commun.* 10 (2019): 2577.

13. Melissa Coleman, email message to author, July 6, 2021.

14. Eric Fortune, email message to author, July 1, 2021.

15. Eliot A. Brenowitz, "Taking turns: The neural control of birdsong duets," *Proc. Natl. Acad. Sci. USA* 118 (2021): e2108043118.

16. Tina C. Roeske, David Rothenberg, and David E. Gammon, "Mockingbird morphing music: Structured transitions in a complex bird song," *Front. Psychol.* 12 (2021): 630115.

17. David Rothenberg, "Mockingbird song decoded," November 3, 2020, YouTube video, 4:53, https://youtu.be/FwD0ij_CWoM.

18. Logan S. James et al., "Phylogeny and mechanisms of shared hierarchical patterns in birdsong," *Curr. Biol.* 31 (2021): 2796–2808.

19. Iris Adam et al., "One-to-one innervation of vocal muscles allows precise control of birdsong," *Curr. Biol.* 31 (2021) 3115–3124.

20. Tomoko G. Fujii, Maki Ikebuchi, and Kazuo Okanoya, "Sex differences in the development and expression of a preference for familiar vocal signals in songbirds," *PLoS ONE* 16 (2021): e0243811.

21. Danielle M. Ferraro et al., "The phantom chorus: Birdsong boosts human well-being in protected areas," *Proc. R. Soc. B* 287 (2020): 20201811.

22. Ellen C. Garland et al., "Song hybridization events during revolutionary song change provide insights into cultural transmission in humpback whales," *Proc. Natl. Acad. Sci. USA* 114 (2017): 7822–7829.

23. Benjamin L. Gottesman et al., "What does resilience sound like? Coral reef and dry forest acoustic communities respond differently to Hurricane Maria," *Ecol. Indic.* 126 (2021): 107635.

24. T. Aran Mooney et al., "Variation in hearing within a wild population of beluga whales (*Delphinapterus leucas*)," *J. Exp. Biol.* 221 (2018): jeb171959.

25. T. Aran Mooney et al., "Local acoustic habitat relative to hearing sensitivities in beluga whales (*Delphinapterus leucas*)," *J. Ecoacoust.* 2 (2018): #QZD9Z5.

26. Jens C. Koblitz et al., "Highly directional sonar beam of narwhals (*Monodon monoceros*) measured with a vertical 16 hydrophone array," *PLoS ONE* 11 (2016): e0162069.

27. Susanna B. Blackwell et al., "Spatial and temporal patterns of sound production in East Greenland narwhals," *PLoS ONE* 13 (2018): e0198295.

28. Danielle Cholewiak et al., "Communicating amidst the noise: Modeling the aggregate influence of ambient and vessel noise on baleen whale communication space in a national marine sanctuary," *Endang. Species. Res.* 36 (2018): 59–75.

29. Hannah B. Blair et al., "Evidence for ship noise impacts on humpback whale foraging behaviour," *Biol. Lett.* 12 (2016): 20160005.
30. K. M. Stafford et al., "Extreme diversity in the songs of Spitsbergen's bowhead whales," *Biol. Lett.* 14 (2018): 20180056.
31. Mia L. K. Nielsen et al., "Acoustic crypsis in southern right whale mother–calf pairs: Infrequent, low-output calls to avoid predation?," *J. Exp. Biol.* 222 (2019): jeb190728.
32. Samuel R. P.-J. Ross et al., "Listening to ecosystems: Data-rich acoustic monitoring through landscape-scale sensor networks," *Ecol. Res.* 33 (2018): 135–147.
33. Samuel R. P.-J. Ross et al., "Utility of acoustic indices for ecological monitoring in complex sonic environments," *Ecol. Indic.* 121 (2021): 107114.
34. Timothy J. Boycott, Jingyi Gao, and Megan D. Gall, "Deer browsing alters sound propagation in temperate deciduous forests," *PLoS ONE* 14 (2019): e0211569.
35. Timothy C. Mullet et al., "Temporal and spatial variation of a winter soundscape in south-central Alaska," *Landscape Ecol.* 31 (2016): 1117–1137.
36. Amandine Gasc et al., "Soundscapes reveal disturbance impacts: Biophonic response to wildfire in the Sonoran Desert Sky Islands," *Landsc. Ecol.* 33 (2018): 1399–1415.
37. Rob Williams et al., "Noise from deep-sea mining may span vast ocean areas," *Science* 377 (2022): 157–158.
38. Patrick J. O. Miller et al., "Behavioral responses to predatory sounds predict sensitivity of cetaceans to anthropogenic noise within a soundscape of fear," *Proc. Natl. Acad. Sci. USA* 119 (2022): e2114932119.
39. P. L. Tyack et al., "Beaked whales respond to simulated and actual navy sonar," *PLoS ONE* 6 (2011): e17009.
40. Terrie M. Williams et al., "Physiological responses of narwhals to anthropogenic noise: A case study with seismic airguns and vessel traffic in the Arctic," *Funct. Ecol.* 36 (2022): 2251–2266.
41. Carlos M. Duarte et al., "The soundscape of the Anthropocene ocean," *Science* 371 (2021): 583.
42. Sophie L. Nedelec et al., "Limiting motorboat noise on coral reefs boosts fish reproductive success," *Nat. Commun.* 13 (2022): 2822.

43.   Andrew Thaler, "Deep-sea mining state of technology, 2022," DSM Observer, April 27, 2022, https://dsmobserver.com/2022/04/deep-sea -mining-state-of-technology-2022/.

CHAPTER 9   Animals Shaping the Environment

1.   Christopher E. Doughty et al., "Global nutrient transport in a world of giants," *Proc. Natl. Acad. Sci. USA* 113 (2016): 868–873.

2.   Anthony D. Barnosky et al., "Variable impact of Late-Quaternary megafaunal extinction in causing ecological state shifts in North and South America," *Proc. Natl. Acad. Sci. USA* 113 (2015): 856–861.

3.   Christopher E. Doughty et al., "Global nutrient transport in a world of giants."

4.   Beth Shapiro, "Could we 'de-extinctify' the woolly mammoth?," *Observer*, April 26, 2015, https://www.theguardian.com/science/2015 /apr/26/woolly-mammoth-normal-for-norfolk-de-extinction.

5.   Kristin Hugo, "George Church on mammoths," *Strange Biology* (blog), October 17, 2015, https://www.strangebio.com/post/131386740854 /george-church-on-mammoths-george-church-is-a?is_related_post=1.

6.   Kenyon B. Mobley et al., "Home ground advantage: Local Atlantic salmon have higher reproductive fitness than dispersers in the wild," *Sci. Adv.* 5 (2019): eaav1112.

7.   Michelle M. Scanlan et al., "Magnetic map in nonanadromous Atlantic salmon," *Proc. Natl. Acad. Sci. USA* 115 (2018): 10995–10999.

8.   Jacques Leslie, "After a long boom, an uncertain future for big dam projects," *Yale Environment 360*, November 27, 2018, https://e360.yale .edu/features/after-a-long-boom-an-uncertain-future-for-big-dam -projects.

9.   Lisa Crozier, email message to author, November 27, 2019.

10.   James R. Faulkner et al., "Associations among fish length, dam passage history, and survival to adulthood in two at-risk species of Pacific salmon," *T. Am. Fish. Soc.* 148 (2019): 1069–1087.

11.   Jack Bloomer, David Sear, and Paul Kemp, "Does variation in egg structure among five populations of Atlantic salmon (*Salmo salar*) influence their survival in low oxygen conditions?," *R. Soc. Open Sci.* 6 (2019): 181020.

12.   Yann Czorlich et al., "Rapid sex-specific evolution of age at maturity is shaped by genetic architecture in Atlantic salmon," *Nat. Ecol. Evol.* 2 (2018): 1800–1807.

13.   Sean R. Brennan et al., "Shifting habitat mosaics and fish production across river basins," *Science* 364 (2019): 783–786.

14.   Lisa G. Crozier et al., "Climate vulnerability assessment for Pacific salmon and steelhead in the California Current Large Marine Ecosystem," *PLoS ONE* 14 (2019): e0217711.

15.   Chase R. Williams et al., "Elevated CO_2 impairs olfactory-mediated neural and behavioral responses and gene expression in ocean-phase coho salmon (*Oncorhynchus kisutch*)," *Glob. Change Biol.* 25 (2019): 963–977.

16.   Camilla Brattland and Tero Mustonen, "How traditional knowledge comes to matter in Atlantic salmon governance in Norway and Finland," *Arctic* 71 (2018): 365–482.

17.   Pablo I. Plaza, Guillermo Blanco, and Sergio A. Lambertucci, "Implications of bacterial, viral and mycotic microorganisms in vultures for wildlife conservation, ecosystem services and public health," *Ibis* 162 (2020): 1109–1124.

18.   Meredith L. Gore et al., "A conservation criminology-based desk assessment of vulture poisoning in the Great Limpopo Transfrontier Conservation Area," *Glob. Ecol. Conserv.* 23 (2020): e01076.

19.   "Vulture MsAP strategic implementation plan launched," CMS, February 18, 2020, https://www.cms.int/raptors/en/news/vulture-msap-strategic-implementation-plan-launched.

20.   Lewis Kihumba, "Investigating the mystery behind Guinea-Bissau's mass vulture deaths," BirdLife International, May 5, 2020, https://www.birdlife.org/news/2020/05/05/investigating-the-mystery-behind-guinea-bissaus-mass-vulture-deaths/.

21.   "New project combats illegal vulture body parts trade in Nigeria," Vulture Conservation Foundation, June 1, 2019, https://4vultures.org/blog/2019-06-01-new-project-combats-illegal-vulture-body-parts-trade-nigeria.

22.   A. J. Botha et al., *Multi-species action plan to conserve African-Eurasian vultures*, CMS Raptors MOU Technical Publication no. 5, CMS Technical Series no. 35 (Abu Dhabi, United Arab Emirates: Coordinating Unit of the CMS Raptors MOU, 2017), https://www.cms.int/raptors/en/publication/multi-species-action-plan-conserve-african-eurasian-vultures-vulture-msap-cms-technical.

23.   D. E. Pritchard, *Strategic implementation plan (2020–2023) for the multi-species action plan to conserve African-Eurasian vultures*

(Vulture MsAP), CMS Raptors MOU Technical Publication No. 7, CMS Technical Series No. 42 (Abu Dhabi, United Arab Emirates: Coordinating Unit of the CMS Raptors MOU, 2020), https://www.cms.int/en /publication/vulture-msap-strategic-implementation-plan-report -implementation-date-cms-technical.

24. Evan R. Buechley et al., "Identifying critical migratory bottlenecks and high-use areas for an endangered migratory soaring bird across three continents," *J. Avian Biol.* 49 (2018): e01629.

25. Evan R. Buechley et al., "Identifying critical migratory bottlenecks."

26. H. J. Williams et al., "Physical limits of flight performance in the heaviest soaring bird," *Proc. Natl. Acad. Sci. USA* 117 (2020): 17884–17890.

CHAPTER 10   Life in the Times of Climate Change

1. Alfred Wegener Institute, "The grand finale to the expedition of a century," news release, October 12, 2020, https://www.awi.de/en/about -us/service/press/single-view/the-grand-finale-to-the-expedition-of-a -century.html.

2. Hotaek Park et al., "Increasing riverine heat influx triggers Arctic sea ice decline and oceanic and atmospheric warming," *Sci. Adv.* 6 (2020): eabc4699.

3. Sarah C. Davidson et al., "Ecological insights from three decades of animal movement tracking across a changing Arctic," *Science* 370 (2020): 712–715.

4. Logan T. Berner et al., "Summer warming explains widespread but not uniform greening in the Arctic tundra biome," *Nat. Commun.* 11 (2020): 4621.

5. K. M. Lewis, G. L. van Dijken, and K. R. Arrigo, "Changes in phytoplankton concentration now drive increased Arctic Ocean primary production," *Science* 369 (2020): 198–202.

6. Jonathan Watts, "Arctic methane deposits 'starting to release', scientists say," *Guardian*, October 27, 2020, https://www.theguardian.com /science/2020/oct/27/sleeping-giant-arctic-methane-deposits-starting -to-release-scientists-find.

7. Huiqi Chen et al., "Spatiotemporal variation of mortality burden attributable to heatwaves in China, 1979–2020," *Sci. Bull.* 67 (2022): 1340–1344.

8. Damian Carrington, "Day of 40C shocks scientists as UK heat record 'absolutely obliterated,'" *Guardian*, July 19, 2022, https://www .theguardian.com/environment/2022/jul/19/day-of-40c-shocks -scientists-as-uk-heat-record-absolutely-obliterated.

9. Ben Clarke et al., "Extreme weather impacts of climate change: An attribution perspective," *Environ. Res.: Climate* 1 (2022): 012001.

10. Luke J. Harrington et al., "Limited role of climate change in extreme low rainfall associated with southern Madagascar food insecurity, 2019–21," *Environ. Res.: Climate* 1 (2022): 021003.

11. IPCC, *Climate change 2023: Synthesis report* (Geneva: IPCC, 2023), https://doi.org/10.59327/IPCC/AR6-9789291691647.

12. Gaia Vince, "The century of climate migration: Why we need to plan for the great upheaval," *Guardian*, August 18, 2022, https://www .theguardian.com/news/2022/aug/18/century-climate-crisis-migration -why-we-need-plan-great-upheaval.

13. Deborah Pardo et al., "Effect of extreme sea surface temperature events on the demography of an age-structured albatross population," *Phil. Trans. R. Soc. B* 372 (2017): 20160143.

14. John C. Wingfield et al., "How birds cope physiologically and behaviourally with extreme climatic events," *Phil. Trans. R. Soc. B* 372 (2017): 20160140.

15. Janet L. Gardner et al., "Effects of extreme weather on two sympatric Australian passerine bird species," *Phil. Trans. R. Soc. B* 372 (2017): 20160148.

16. Pascal Marrot, Dany Garant, and Anne Charmantier, "Multiple extreme climatic events strengthen selection for earlier breeding in a wild passerine," *Phil. Trans. R. Soc. B* 372 (2017): 20160372.

17. BirdLife International, *State of the world's birds 2022: Insights and solutions for the biodiversity crisis* (Cambridge, UK: BirdLife International, 2022), https://www.birdlife.org/papers-reports/state-of-the -worlds-birds-2022/.

18. National Audubon Society, "Survival by degrees: 389 bird species on the brink," accessed December 28, 2023, https://www.audubon.org /climate/survivalbydegrees.

19. Clark S. Rushing et al., "Migratory behavior and winter geography drive differential range shifts of eastern birds in response to recent climate change," *Proc. Natl. Acad. Sci. USA* 117 (2020): 12897–12903.

20. BirdLife International, *State of the world's birds 2022*.

21. Benjamin G. Freeman et al., "Climate change causes upslope shifts and mountaintop extirpations in a tropical bird community," *Proc. Natl. Acad. Sci. USA* 115 (2018): 11982–11987.
22. Thomas K. Lameris et al., "Arctic geese tune migration to a warming climate but still suffer from a phenological mismatch," *Curr. Biol.* 28 (2018): 2467–2473.
23. Brooke Bateman, "False springs: How earlier spring with climate change wreaks havoc on birds," National Audubon Society, September 25, 2020, https://www.audubon.org/news/false-springs-how-earlier-spring-climate-change-wreaks-havoc-birds.
24. Justin R. Eastwood et al., "Hot and dry conditions predict shorter nestling telomeres in an endangered songbird: Implications for population persistence," *Proc. Natl. Acad. Sci. USA* 119 (2022): e2122944119.
25. David López-Idiáquez et al., "Long-term decrease in coloration: A consequence of climate change?," *Am. Nat.* 200 (2022): 32–47.
26. University of the Basque Country, "Research suggests that change in bird coloration is due to climate change," Phys.org, July 14, 2022, https://phys.org/news/2022-07-bird-due-climate.html.
27. Nicholas B. Pattinson et al., "Collapse of breeding success in desert-dwelling hornbills evident within a single decade," *Front. Ecol. Evol.* 10 (2022), https://doi.org/10.3389/fevo.2022.842264.

CHAPTER 11   Our Shared Burden of Disease

1. Jared Diamond, *Guns, germs, and steel: The fates of human societies* (New York: W. W. Norton & Company, 1997).
2. "WHO COVID-19 dashboard," World Health Organization, accessed January 14, 2024, https://data.who.int/dashboards/covid19/deaths.
3. Livia V. Patrono et al., "Archival influenza virus genomes from Europe reveal genomic variability during the 1918 pandemic," *Nat. Commun.* 13 (2022): 2314.
4. Maria A. Spyrou et al., "The source of the Black Death in fourteenth-century central Eurasia," *Nature* 606 (2022): 718–724.
5. Max Planck Institute for Evolutionary Anthropology, "Origins of the Black Death identified," EurekAlert! news release, June 15, 2022, https://www.eurekalert.org/news-releases/955621.
6. A. Izdebski et al., "Palaeoecological data indicates land-use changes across Europe linked to spatial heterogeneity in mortality during the Black Death pandemic," *Nat. Ecol. Evol.* 6 (2022): 297–306.

7. Peter Sarris, "Viewpoint new approaches to the 'Plague of Justinian,'" *Past & Present* 254 (2022): 315–346.

8. Lee Mordechai et al., "The Justinianic Plague: An inconsequential pandemic?," *Proc. Natl. Acad. Sci. USA* 116 (2019): 25546–25554.

9. Sarris, "Viewpoint new approaches to the 'Plague of Justinian,'" 345.

10. Gunnar U. Neumann et al., "Ancient *Yersinia pestis* and *Salmonella enterica* genomes from Bronze Age Crete," *Curr. Biol.* 32 (2022): 3641–3649.

11. Aida Andrades Valtueña et al., "Stone Age *Yersinia pestis* genomes shed light on the early evolution, diversity, and ecology of plague," *Proc. Natl. Acad. Sci. USA* 119 (2022): e2116722119.

12. Meriam Guellil et al., "Ancient herpes simplex 1 genomes reveal recent viral structure in Eurasia," *Sci Adv.* 8 (2022): abo4435; Meriam Guellil et al., "An invasive *Haemophilus influenzae* serotype b infection in an Anglo-Saxon plague victim," *Genome Biol.* 23 (2022): 22.

13. Ben Hu et al., "Discovery of a rich gene pool of bat SARS-related coronaviruses provides new insights into the origin of SARS coronavirus," *PLoS Pathog.* 13 (2017): e1006698.

14. Cara E. Brook et al., "Accelerated viral dynamics in bat cell lines, with implications for zoonotic emergence," *eLife* 9 (2020): e48401.

15. Jian Zheng et al., "Middle East respiratory syndrome coronavirus seropositivity in camel handlers and their families, Pakistan," *Emerg. Infect. Dis.* 25 (2019): 2307–2309.

16. Elmoubasher Farag et al., "MERS-CoV in camels but not camel handlers, Sudan, 2015 and 2017," *Emerg. Infect. Dis.* 25 (2019): 2333–2335.

17. Emmie de Wit et al., "Prophylactic and therapeutic remdesivir (GS-5734) treatment in the rhesus macaque model of MERS-CoV infection," *Proc. Natl. Acad. Sci. USA* 117 (2020): 6771–6776.

18. Li-Li Ren et al., "Identification of a novel coronavirus causing severe pneumonia in human: A descriptive study," *Chin. Med. J.* 133 (2020): 1015–1024.

19. Ping Liu, Wu Chen, and Jin-Ping Chen, "Viral metagenomics revealed Sendai virus and coronavirus infection of Malayan pangolins (*Manis javanica*)," *Viruses* 11 (2019): 979.

20. Kangpeng Xiao et al., "Isolation of SARS-CoV-2-related coronavirus from Malayan pangolins," *Nature* 583 (2020): 286–289.

21. Michael Safi and Eli Block, "'Being truthful is essential': Scientist who stumbled upon Wuhan Covid data speaks out," *Guardian*, March 28,

2023, https://www.theguardian.com/world/2023/mar/28/being -truthful-is-essential-scientist-who-stumbled-upon-wuhan-covid-data -speaks-out.

22. Thomas Kruse et al., "The Ebola virus nucleoprotein recruits the host PP2A-b56 phosphatase to activate transcriptional support activity of VP30," *Mol. Cell* 69 (2018): 136–145.

23. Ana Maria Henao-Restrepo, "Efficacy and effectiveness of an rVSV-vectored vaccine in preventing Ebola virus disease: Final results from the Guinea ring vaccination, open-label, cluster-randomised trial (*Ebola ça suffit!*)," *Lancet* 389 (2017): 505–518.

24. Jesús Olivero et al., "Recent loss of closed forests is associated with Ebola virus disease outbreaks," *Sci. Rep.* 7 (2017): 14291.

25. Anne W. Rimoin et al., "Ebola virus neutralizing antibodies detectable in survivors of the Yambuku, Zaire outbreak 40 years after infection," *J. Infect. Dis.* 217 (2017): 223–231.

CHAPTER 12   The Anthropocene and Beyond

1. HydroLAKES v. 1.0, https://www.hydrosheds.org/products /hydrolakes.

2. Mathis Loïc Messager et al., "Estimating the volume and age of water stored in global lakes using a geo-statistical approach," *Nat. Commun.* 7 (2016): 13603.

3. Miquel Lürling et al., "A critical perspective on geo-engineering for eutrophication management in lakes," *Water Res.* 97 (2016): 1–10.

4. Matthew J. Hoffman and Eric Hittinger, "Inventory and transport of plastic debris in the Laurentian Great Lakes," *Mar. Pollut. Bull.* 115 (2017): 273–281.

5. Anne A. van Dam et al., "Linking hydrology, ecosystem function, and livelihood outcomes in African papyrus wetlands using a Bayesian network model," *Wetlands* 33 (2013): 381–397.

6. Michael Gross, "Ebb tidings," *Curr. Biol.* 16 (2006): R266–R267.

7. David Harper, email message to author, January 2017.

8. Food and Agriculture Organization of the United Nations, *Status of the world's soil resources: Main report* (Rome, Italy: Food and Agriculture Organization of the United Nations and Intergovernmental Technical Panel on Soils, 2015), https://www.fao.org/policy-support /tools-and-publications/resources-details/en/c/435200/.

9.   Anthony Foucher et al., "Increase in soil erosion after agricultural intensification: Evidence from a lowland basin in France," *Anthropocene* 7 (2014): 30–41

10.   Food and Agriculture Organization, "Soils are endangered, but the degradation can be rolled back," April 12, 2015, https://www.fao.org/newsroom/detail/Soils-are-endangered-but-the-degradation-can-be-rolled-back/.

11.   Food and Agriculture Organization, *Status of the world's soil resources.*

12.   E. W. Slessarev et al., "Water balance creates a threshold in soil pH at the global scale," *Nature* 540 (2016): 567–569.

13.   J. H. Guo et al., "Significant acidification in major Chinese croplands," *Science* 327 (2010): 1008–1010.

14.   François P. Teste et al., "Plant-soil feedback and the maintenance of diversity in Mediterranean-climate shrublands," *Science* 355 (2017): 173–176.

15.   Jonathan A. Bennett et al., "Plant-soil feedbacks and mycorrhizal type influence temperate forest population dynamics," *Science* 355 (2017): 181–184.

16.   Wim H. van der Putten, "Belowground drivers of plant diversity," *Science* 355 (2017): 134–135.

17.   Food and Agriculture Organization, "Soils are endangered."

18.   Jennifer L. Lavers and Alexander L. Bond, "Exceptional and rapid accumulation of anthropogenic debris on one of the world's most remote and pristine islands," *Proc. Natl. Acad. Sci. USA* 114 (2017): 6052–6055.

19.   Jenna R. Jambeck et al., "Plastic waste inputs from land into the ocean," *Science* 347 (2015): 768–771.

20.   Roland Geyer, Jenna R. Jambeck, and Kara Lavender Law, "Production, use, and fate of all plastics ever made," *Sci. Adv.* 3 (2017): e1700782.

21.   Chris Jordan, *Midway: Message from the gyre,* accessed March 30, 2023, http://www.chrisjordan.com/gallery/midway/.

22.   Matthew S. Savoca et al., "Marine plastic debris emits a keystone infochemical for olfactory foraging seabirds," *Sci. Adv.* 2 (2016): e1600395.

23.   Jan Zalasiewicz et al., "The geological cycle of plastics and their use as a stratigraphic indicator of the Anthropocene," *Anthropocene* 13 (2016): 4–17.

24. Jan Zalasiewicz et al., "Scale and diversity of the physical techno-sphere: A geological perspective," *Anthr. Rev.* 4 (2016): 9–22.

25. Foresight, Research and Innovation, *2050 Scenarios: Four plausible futures*, ARUP, December 2019, https://www.arup.com/perspectives/publications/research/section/2050-scenarios-four-plausible-futures.

26. Jan Zalasiewicz, email message to author, December 10, 2019.

27. Maggie J. Watson and David M. Watson, "Post-Anthropocene conservation," *Trends Ecol. Evol.* 35 (2020): 1–3.

28. Fernando de Sousa Mello and Amâncio César Santos Friaça, "The end of life on Earth is not the end of the world: Converging to an estimate of life span of the biosphere?," *Int. J. Astrobiol.* 19 (2019): 25–42.

29. Jack T. O'Malley-James et al., "Swansong Biospheres: Refuges for life and novel microbial biospheres on terrestrial planets near the end of their habitable lifetimes," *Int. J. Astrobiol.*, 12 (2013): 99–112.

30. Watson and Watson, "Post-Anthropocene conservation."

31. Watson and Watson, "Post-Anthropocene conservation."

INDEX

Main entries use systematic names when the common names are too complex (e.g., black-throated blue warbler) or too obscure or when different names are used for the same species (e.g., caribou and reindeer).

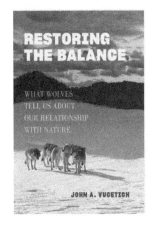
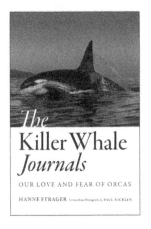

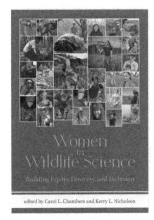